S247 INORGANIC CHEMISTRY
CONCEPTS AND CASE STUDIES
SCIENCE: A SECOND LEVEL COURSE

BLOCK 4
MOLECULES GREAT AND SMALL

PREPARED FOR THE COURSE TEAM BY ELAINE MOORE

THE OPEN UNIVERSITY

S247 REWRITE COURSE TEAM

CHAIR
David Johnson (1989–90)
Lesley E. Smart (1990–)

AUTHORS
David Johnson
Elaine A. Moore
Jane Nelson
Lesley E. Smart

COURSE COORDINATOR
Wendy L. Selina

EDITOR
Dick Sharp

BBC
Barrie Whatley

DESIGN GROUP
Debbie Crouch (designer)
Alison George (graphic artist)
Roy Lawrance (graphic artist)

SECRETARIAL SUPPORT
Jenny Burrage
Sally Eaton
Sue Hegarty

ORIGINAL S247 COURSE TEAM

CHAIR
Stuart Bennett

AUTHORS
Stuart Bennett (Blocks 2 and 3; Case Study 4)
Charles Harding (Block 6)
David Johnson (Block 1; Case Study 1)
Joan Mason (Block 5)
Elaine A. Moore (Block 4)
Jane Nelson (Block 7)
Lesley E. Smart (Blocks 2 and 7)
Keith Trigwell (Case Study 2)
Kiki Warr (Block 1)
John Emsley (Block 7; Case Study 4)
Harry Kroto (Case Study 3)

COURSE COORDINATOR
Keith Trigwell

The Open University, Walton Hall, Milton Keynes, MK7 6AA.

First published 1981.

Second edition 1983.

Third edition 1993.

Reprinted 1996, 1999 (with corrections).

Edited, designed and typeset by The Open University.

Printed in the United Kingdom by Thanet Press Limited, Margate, Kent.

ISBN 07492 50933

This text forms part of an Open University Level 2 course. If you would like a copy of *Studying with the Open University,* please write to the Course Reservations and Sales Centre, PO Box 724, The Open University, Walton Hall, Milton Keynes, MK7 6ZS, United Kingdom. If you have not enrolled on the Course and would like to buy this or other Open University material, please write to Open University Worldwide, The Berrill Building, Walton Hall, Milton Keynes, MK7 6AA.

3.3

19080C/s247b4i3.3

VALENCE-SHELL ELECTRON-PAIR REPULSION THEORY

1 Draw a Lewis structure of the molecule, and assign valence electrons to covalent (including dative) bonds as far as possible.

2 As far as possible, arrange into lone pairs the non-bonded valence-shell electrons around the atoms that are bound to more than one other atom.

3 Each bond, each lone pair, and each odd electron around a particular atom is a repulsion axis. A double or triple bond is only one bond for this purpose.

4 Arrange the repulsion axes to be as far apart as possible.

5 Assign non-equivalent axes to possible sites, using the following orders of electron-pair repulsion energies:

(a) lone pair : lone pair > lone pair : bond pair > bond pair : bond pair >odd electron > electron pair;

(b) triple bond > double bond > single bond.

6 Use the order of interactions in step 5 to decide whether bond angles should be larger or smaller than the regular angle.

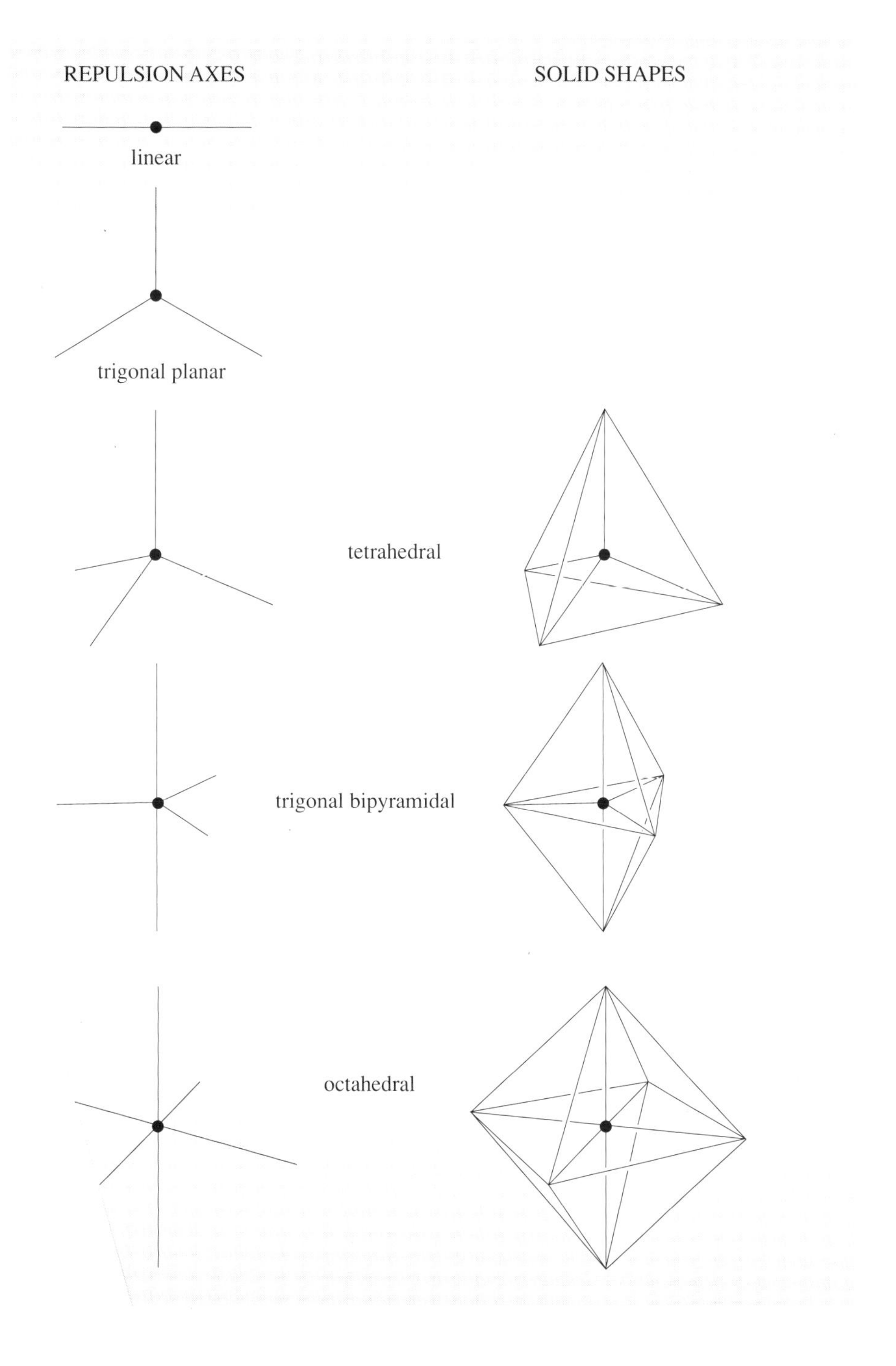

CONTENTS

STUDY GUIDE FOR BLOCK 4

The components of this Block are the main text and three video sequences. The Block should occupy three study weeks. Case Study 2, which is associated with this Block, should take a further half a study week. There are two more video sequences to go with the Case Study.

During the first study week you should aim to read up to the end of Section 5. Video sequences 5 and 6 are designed to aid your understanding of symmetry; you should view sequence 5 just before or during your study of Section 5, and sequence 6 afterwards. Sections 6 and 7 should occupy the second study week, and the third video sequence (sequence 7) is related to these sections. The third study week is devoted to finishing Section 7 and reading Section 8. The ideas in these later Sections are quite sophisticated, and if you run short of time you may wish to read Sections 8.3 and 8.4 fairly quickly before moving on to Section 8.5 and the Case Study. Video sequences 8 and 9 should be viewed before or during your reading of Case Study 2.

There are no Home Experiments associated with this Block, but your Orbit model kit will help you to visualize the molecular shapes discussed in Sections 3–6.

1 INTRODUCTION

'...and now for something completely different.'

So far in this Course we have dealt mainly with metals and their compounds. In later Blocks, you will study aspects of the chemistry of non-metals. In this Block, we introduce you to theories of chemical bonding that will be especially helpful in your study of non-metal compounds.

The thermodynamics that you studied in the earlier Blocks tell us whether a reaction is possible or not. It may tell us that a certain molecule is stable mainly because the atoms are bound together very strongly. It does not tell us why that particular molecule should be so strongly bound. This is the realm of bonding theories.

Theories of chemical bonding are concerned with the way in which atoms combine to form molecules. You will meet two bonding theories in this Block. The first is the simple electron-pair model that you met in the Science Foundation Course. The second is a more sophisticated theory, molecular orbital theory, which has proved to be capable of explaining certain molecular properties that the elementary theory could not. For example, why do liquid oxygen and liquid nitrogen behave differently in a magnetic field?

Molecular orbital theory is the preferred theory of chemists who calculate properties of molecules. It is used to calculate not only bonding energies but also other properties of molecules, such as bond lengths and how the electron charge is distributed over the atoms, and it can even be used to picture the way molecules change when they react.

Some of the properties that molecular orbital theory can explain are illustrated on the videocassette. In this we also try to help you to become familiar with the idea of electrons behaving like waves, because this wave-like behaviour is the basis of molecular orbital theory.

In Block 6 you will meet some methods of determining molecular shapes using spectroscopy. In this Block we show how simple bonding theory can be used to predict the shapes of molecules. We also introduce you to a way of classifying shapes. One of the things you probably noticed about the crystals in Block 2 was how symmetrical they were. Here we build on intuitive ideas of what is symmetrical to develop a formal description of molecular shape. This description is introduced on the videocassette and developed further in the text. In this Block it will be used to simplify molecular orbital theory for molecules with three or more atoms, and in Block 6 you will meet it again when we look at the spectra of these molecules.

In most of this Block we shall be looking at small molecules, in particular those containing only two atoms (diatomic molecules). In the following Case Study, however, we shall see that we can apply molecular orbital theory to some very large 'molecules' – crystals of certain solids. Application of molecular orbital theory to crystals of semi-metals gives us some insight into the properties of semiconductors – materials widely used in computers, microprocessors (silicon chips), radios, photocopying machines, etc. The use of semiconductors in solar cells will be discussed in videosequence 8, which will also introduce Case Study 2. The Case Study looks in detail at how semiconductors are used to produce electricity from sunlight.

As it is easier to see the shape of a molecule from a model than from a drawing, we have suggested that you construct models of several molecules. We advise you, therefore, to have your Orbit model kit with you when you read this Block.

2 LEWIS STRUCTURES AND ELECTRON-PAIR BONDS

The observation of the lack of reactivity of the noble gases (He, Ne, Ar, Kr and Xe) led to the proposal that the electron configurations of their atoms were particularly stable and that the atoms of other elements strove to achieve these configurations by forming compounds.

Apart from helium (He $1s^2$), the outer electron configurations of the noble gases can be expressed as ns^2np^6. Thus the outermost shell contains eight electrons. These eight electrons are known as an **octet**.

A very simple bonding theory based on these observations can be stated thus: when elements combine to form compounds, the atoms of each element lose or gain electrons to form an octet. For example, an oxygen atom has the electron configuration $1s^22s^22p^4$. It is therefore two electrons short of an octet in its outer shell (n = 2) and, when oxygen combines to form compounds, the oxygen atoms in the compounds will have each gained two electrons. Sodium, to take an example from the other side of the Periodic Table, has the configuration $1s^22s^22p^63s^1$, with only one electron in its outer shell (n = 3). In this case the easier way to achieve an octet is to lose an electron.

■ Write down the electron configurations of (a) a magnesium atom and (b) a nitrogen atom. How will these two elements most easily achieve an octet?

□ (a) Magnesium has the configuration $1s^22s^22p^63s^2$; it will lose two electrons. (b) Nitrogen has the configuration $1s^22s^22p^3$; it most easily achieves an octet by gaining three electrons.

There are two ways in which an atom can gain electrons to form an octet. It can form an ion, such as O^{2-}, or it can share electrons with another atom. For example, fluorine has the configuration $1s^22s^22p^5$ and needs to gain one electron to make up its octet. It can do this by forming fluoride ions, F^-, as in the salt sodium fluoride, NaF. Alternatively, it can acquire a share of an electron belonging to another atom. In difluorine, F_2, for example, each fluorine attains an octet by borrowing an electron from the other fluorine.

When atoms form ions, the compounds are held together by the electrostatic interaction of the oppositely charged ions. This is called ionic bonding.

The sharing of electrons between atoms is called covalent bonding. Each pair of electrons that is shared between two atoms forms one covalent bond. In difluorine there are two shared electrons, one pair, and thus one bond. F_2 is said to have a single bond. In dioxygen, O_2, the oxygen atoms share two pairs of electrons and so O_2 has a double bond.

The electron configurations of atoms and the completion of octets upon formation of molecules can be shown diagrammatically by the use of **Lewis structures**. In Lewis structures, the outermost or valence electrons are represented by small circles and crosses: a sodium atom, for example, is written $Na^{\times}$ and a fluorine atom is written 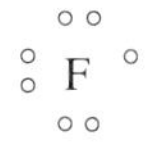. If these atoms now lose and gain electrons to form ions, the Lewis structures become $[Na]^{+}$ and 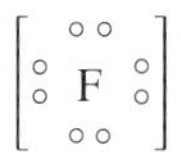. (It is usual to show only the electrons in the original outer shell, so in the example just given, where Na loses one electron we do not put eight dots or crosses around Na to show the $n = 2$ shell but leave it blank to indicate loss of the one electron from the $n = 3$ shell.)

When, as in difluorine, two electrons are shared, we indicate this by putting these electrons between the two atoms:

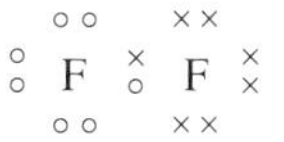

In counting the electrons round each atom, these shared electrons count twice, so that the left-hand fluorine has eight electrons around it and so does the right-hand fluorine, as shown in the margin. To see if you have followed this description of Lewis structures, try some examples for yourself.

■ What are the Lewis structures for (a) ClF, (b) CO_2, (c) H_2O, (d) HCN?

□ (a), (b), (c), (d)

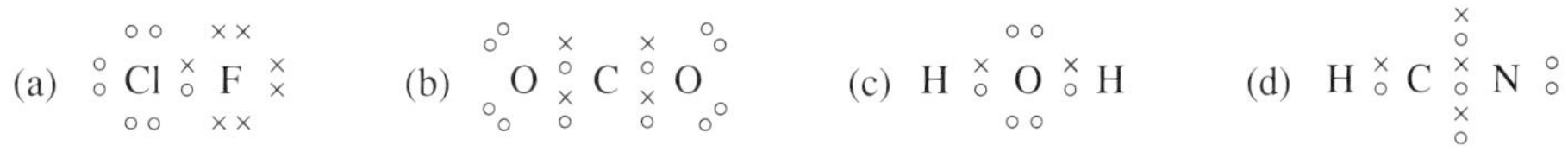

Note that, in CO_2, each oxygen needs to gain two electrons and the carbon needs four electrons. This is achieved by sharing four electrons between the carbon and each oxygen. Four electrons make two pairs of electrons, and so the bonds in CO_2 are double bonds. Similarly, in HCN, three electron pairs are shared between C and N, giving a triple bond.

In Lewis structures, we show covalent bonds by putting the circles and crosses representing the shared electrons between the bonded atoms. A more elegant way of illustrating covalent bonding is to draw a structural formula in which each bond is represented by a line joining the bonded atoms. For the four examples in the question above, the structural formulae are

F—Cl, O=C=O, H—O—H, and H—C≡N

It is sometimes important to indicate outer electrons that are not involved in bonding. These electrons do not bind the atoms together, but because they occupy space, they can affect the shape of the molecule. This is especially true of the non-bonding electrons on atoms bonded to two or more other atoms, such as the O in H_2O. A pair of electrons that is associated with only one atom in a molecule is known as a **lone pair** and can be indicated on structural formulae by two dots, for example

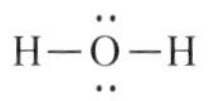

Note that the structural formula for CO_2 remains O=C=O, because all the electrons in the octet around carbon are used for bonding.

■ Write Lewis structures and structural formulae for the following molecules: (a) OF_2; (b) NH_3; (c) HONO (in this molecule the H is attached to an oxygen atom but not to the nitrogen). In the structural formulae indicate the lone pairs on the central atom.

□

Lewis structure	Structural formula
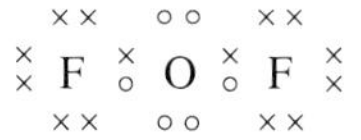 F ×° O °× F (with lone pairs: × on F, ° on O)	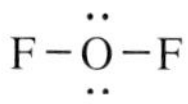 $F{-}\ddot{\underset{..}{O}}{-}F$
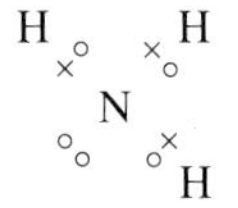 H °× N ×° H, with third H (lone pair ° ° on N)	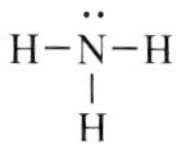 $H{-}\ddot{N}{-}H$ with H below N
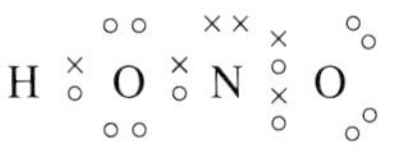 H ×° O ×° N °× O (with lone pairs)	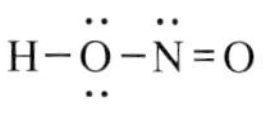 $H{-}\ddot{\underset{..}{O}}{-}\ddot{N}{=}O$

The molecules you have met so far have all been uncharged, but we can equally well draw Lewis structures and structural formulae for ions, such as the carbonate ion, CO_3^{2-}, or the ammonium ion, NH_4^+. There is only one slight complication: the charge is associated with the whole ion and, in order to write Lewis structures, we need to put the extra electrons on or take electrons away from particular atoms. For most purposes, when predicting shapes for example, it does not matter where you add electrons to or take them from, but it probably helps if you adopt a systematic method. One such method is given below.

Put the extra electrons on or remove electrons from the central atom. Thus you would put two electrons round carbon in CO_3^{2-} and remove one from nitrogen in NH_4^+.

We shall show you here how to apply this method to NH_4^+ and CO_3^{2-}. For NH_4^+, we remove an electron from N to give N^+ and then put four H atoms around the N^+. Nitrogen has an outer electron configuration $2s^22p^3$, and so N^+ has only four outer electrons. The Lewis structure of NH_4^+ is thus

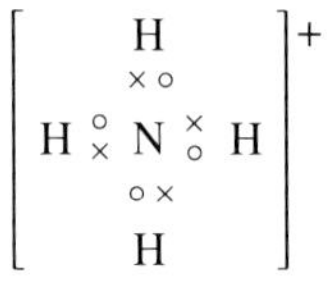

$$\left[\begin{array}{ccc} & H & \\ & \times\circ & \\ H\ {}^{\circ}_{\times} & N & {}^{\times}_{\circ}\ H \\ & \circ\times & \\ & H & \end{array}\right]^+$$

and the structural formula is

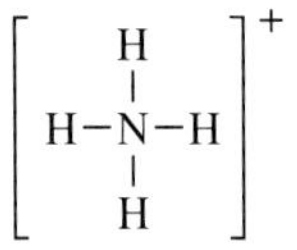

$$\left[\begin{array}{ccc} & H & \\ & | & \\ H- & N & -H \\ & | & \\ & H & \end{array}\right]^+$$

Now let's take CO_3^{2-}. We put two extra electrons on carbon, giving it six outer electrons. If we are to retain an octet around carbon, then we cannot form three carbon–oxygen double bonds as this would cause carbon to be surrounded by twelve outer electrons. If we form one carbon–oxygen double bond then there will be an octet around carbon, but how then can we form the other two carbon–oxygen bonds? Well, they are formed by carbon giving a share of two electrons to oxygen. Such single bonds, where both electrons in the pair originate from one atom, are called **dative bonds**. The Lewis structure of CO_3^{2-} is thus

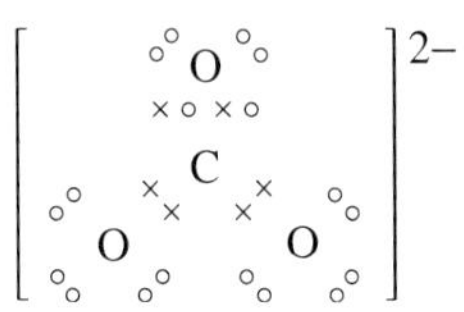

$$\left[\begin{array}{ccc} & {}^{\circ}_{\circ}O{}^{\circ}_{\circ} & \\ & \times\circ\times\circ & \\ & C & \\ {}^{\circ}_{\circ}\ {}^{\times}_{\times} & & {}^{\times}_{\times}\ {}^{\circ}_{\circ} \\ O & & O \\ {}^{\circ}_{\circ}\ {}^{\circ}_{\circ} & & {}^{\circ}_{\circ}\ {}^{\circ}_{\circ} \end{array}\right]^{2-}$$

A dative bond is often denoted by an arrow rather than a line, giving a structural formula like this:

[O=C(→O)(→O)]$^{2-}$

Try a few examples now. The question below includes ions and molecules, some of which require dative bonds.

■ Write Lewis structures and structural formulae for (a) NO_3^-, (b) OH^-, (c) F_3NO (N central atom), and (d) H_3O^+ (O central atom).

□

Lewis structure	Structural formula
[Lewis structure of NO_3^-]	[O=N(→O)(→O)]$^-$
[Lewis structure of OH^-]	[O–H]$^-$
[Lewis structure of F_3NO]	F–N(F)(F)→O
[Lewis structure of H_3O^+]	[H–Ö(H)(H)]$^+$

There is a further refinement to the structural formula that can be made. You may have noticed that in the nitrate anion, NO_3^-, for example, the three oxygens are not bound in exactly the same way to nitrogen – one is doubly bonded and the other two are singly bonded. If, however, we measure the N—O bond lengths in NO_3^-, we find that all three N—O bonds are equivalent. To overcome such difficulties it was proposed that, in such cases, the molecule was not adequately represented by one structural formula but was better represented by a mixture of structural formulae, known as a **resonance hybrid**. Taking NO_3^- as an example, we can represent the anion by the following mixture of structural formulae:

[O=N(→O)(→O)]$^-$ ⟷ [O←N(=O)(→O)]$^-$ ⟷ [O←N(→O)(=O)]$^-$

Such a representation means that each N—O bond is a mixture of one-third double bond and two-thirds single dative bond.

It is important to note, however, that when resonance structures are used, the molecule does *not* change from one form to another; it is a hybrid of the forms shown (in the same way that a mule does not change from a horse to a donkey and back but is a different species from either that is a hybrid of the two).

If a resonance structure involves a bond being a hybrid of double and single bonds as in NO_3^- above, there is an alternative representation in which a dotted line is used for a partial bond. The structural formula of NO_3^- then becomes

indicating that all three N—O bonds are equivalent and between single and double bonds.

Probably the most famous example of a resonance hybrid is the structural formula of the organic molecule benzene, C_6H_6. This is

or, in the representation just introduced,

2.1 BEYOND THE OCTET

Although there are many compounds in which each atom attains an octet, it became apparent that it is not necessary for every atom to have an octet in its outer shell in order for a molecule to be stable. Consider for example the molecule BF_3. Boron has the electron configuration $1s^22s^22p^1$, and in BF_3 the three outer electrons can be used to form single bonds to fluorine:

This, however, leaves boron with only six electrons in its outer shell, not eight. BF_3 is an example of a molecule with too few electrons to form octets around every atom. Such molecules are sometimes known as electron deficient.

IF_7 on the other hand has too many electrons. The central iodine has an outer configuration $5s^25p^5$, with seven electrons. It can form seven covalent bonds thus completing the octets of the fluorine atoms, but it then has fourteen electrons in its outer shell. We cannot draw dative bonds with iodine donating two electrons to fluorine, as fluorine needs only one electron to complete its octet. We are therefore stuck with

This expansion of the octet is found particularly for the larger non-metals, such as phosphorus, sulphur, chlorine, bromine and iodine.

Such molecules force us to revise our theory a bit. We can now say that atoms form stable covalent molecules by using their outer electrons to form electron-pair bonds to other atoms. Atoms in the second row of the Periodic Table (from lithium to neon) generally achieve an octet (although in some cases they have insufficient electrons to do so). Atoms from later rows may contain more than eight electrons in their outer shell when they form a molecule. This expansion of the octet in later rows can be rationalized in terms of the availability of d energy levels. Along the second row the 2s and 2p levels are being filled. There are no 2d electrons. For later rows, the *n*s, *n*p and *n*d energy levels are available. When you are deciding how to draw Lewis structures, try to make an octet around second row atoms, using dative bonds if necessary to avoid having more than eight electrons. With third and later row atoms, you can expand the number of electrons in the outer shell (up to a maximum of 18) and retain simple single and double bonds in preference to dative bonds.

SAQ 1 What are the Lewis structures for the following fluorides: (a) NF_3; (b) PF_5; (c) SF_4; (d) SF_6? In each case, the fluorine atoms surround and are bonded to a central atom.

SAQ 2 Write Lewis structures for the following molecules: (a) hypochlorous acid, HOCl; (b) boron hydrogen difluoride, HBF_2; (c) nitrosyl fluoride, ONF; (d) iodine pentafluoride, IF_5.

SAQ 3 Write structural formulae for the molecules in SAQ 2. Include lone pairs on the atoms as dots.

SAQ 4 Solid aluminium chloride ($AlCl_3$) at room temperature and pressure has a layered structure in which aluminium ions, Al^{3+}, occupy one third of the octahedral holes in a close-packed arrangement of chloride ions. By heating the solid, however, we can obtain aluminium chloride in the vapour phase. The vapour is found to consist of Al_2Cl_6 molecules in which the atoms are arranged as shown below:

```
Cl      Cl      Cl
  \    /  \    /
   Al      Al
  /    \  /    \
Cl      Cl      Cl
```

These molecules are also found in molten aluminium trichloride. Write a structural formula, including lone pairs, for this molecule. (The lines in the diagram above show which atoms are joined and do not imply single covalent bonding.)

SAQ 5 Write structural formulae for the following molecules, indicating lone pairs where appropriate: (a) methanal, H_2CO; (b) diimide, N_2H_2 (HNNH); (c) sulphur trioxide, SO_3; (d) phosphine, PH_3; (e) the phosphine–borane adduct, H_3PBH_3; (f) dinitrogen pentoxide, N_2O_5 (O_2NONO_2).

3 WHAT SHAPES ARE MOLECULES?

Figure 1 shows the experimentally determined shapes of some fluorides. Most of these compounds are stable with respect to the free atoms at room temperature and atmospheric pressure (though some are very reactive). Some contain the molecules shown in the gas phase: BF_3, CF_4, NF_3, PF_5, OF_2, F_2, SiF_4, SF_4, SF_6, and ClF_3 (b.t. 285 K); some in the liquid phase: BrF_5 and IF_7; and some in the solid phase: KrF_2 and XeF_4. You may be surprised by some of them though. Beryllium difluoride at room temperature and atmospheric pressure is, as you may have expected, a crystalline solid. It has a diamond-like structure with each beryllium surrounded by four fluorines. If this solid is heated, however, BeF_2 molecules can be observed in the vapour, and it is this molecule that is shown in Figure 1. The other fluorides that you probably found surprising are the noble gas fluorides KrF_2 and XeF_4. Despite our earlier statement that atoms tend to form bonds in order to attain a noble gas configuration, there is now quite an extensive chemistry of the noble gases, particularly xenon. Noble gas chemistry will be discussed in Block 5. In the modern bonding theory you will meet later in this Block, the attainment of a noble gas configuration no longer plays a part in explaining the formation of covalent molecules, and compounds of the noble gases are treated in the same way as other compounds.

We have chosen to look at fluorides in this Section because they illustrate most of the arrangements of atoms found in covalent molecules. Let's have a look at these shapes. Use your Orbit model kit to make up models of the fluorides shown in Figure 1 and keep these with you when reading Sections 3–6. We found that we needed the following parts from the model kit to build all the fluorides shown, except IF_7:

48 green straws

49 univalent centres for F (pale green)

2 linear divalent centres (1 black for Be, 1 white for Kr)

1 trivalent centre for B (black)

3 tetrahedral centres (1 black for C, 1 silver for Si, 1 blue for N)

1 divalent centre with a bond angle of 110° for O (red)

3 five-valent centres (1 purple for P, 1 red for S, 1 blue for Cl)

3 octahedral centres (1 yellow for S, 1 green for Br, 1 black for Xe)

You will find that you cannot build a model with exactly the angles shown in all cases. Use 3.5 cm green straws to form the bonds. Use a blue tetrahedral centre for nitrogen to make the NF_3 model. In OF_2 the oxygen can be represented by a red divalent centre marked 110. The five-valent atom centres should be used for SF_4 and ClF_3, as well as PF_5. Use octahedral atom centres for SF_6, BrF_5 and XeF_4. Unfortunately, you will not be able to make IF_7, but you should be able to make models of all the other fluorides in Figure 1. IF_7 is like PF_5 but with five bonds round the centre instead of three. A photograph of the completed models (except IF_7) is shown in Colour Plate 1 at the end of the Block.

Let's consider the difluorides. Beryllium difluoride and krypton difluoride are both linear molecules.

■ Which difluoride shown in Figure 1 has a different shape?

□ OF_2 is a bent or V-shaped molecule.

These two shapes (bent and linear) are the only two available to difluorides, and to triatomic molecules generally, although the bond angle can vary.

What sorts of triatomic molecule are linear, and what sorts are bent or V-shaped? Molecules of formula MX_2 occur for the hydrides and oxides of Group VI (O, S, Se, etc.), for the halides of these elements and those of Group II (Be, Mg, etc.), and for the

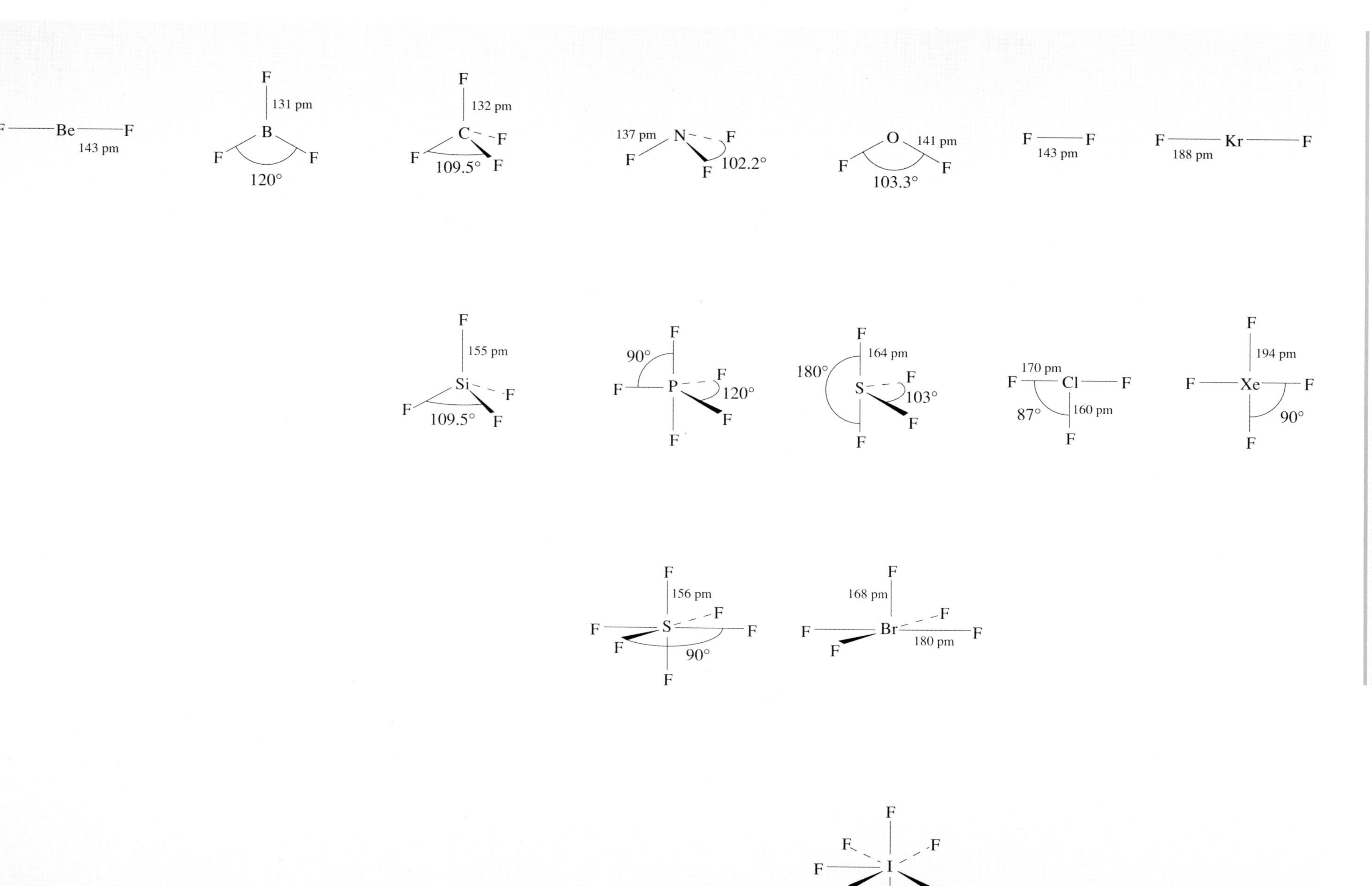

Figure 1 The shapes and sizes of some covalent fluorides.

oxides and sulphides of Group IV (C, Si, etc.). The Group VI hydrides (for example H_2O) are bent, as are their dihalides (see OF_2) and oxides (for example SO_2). The Group II halides vary, the molecules tending to bend more as we go down the group from beryllium, Be, to barium, Ba. The only oxide of Group IV that occurs as a discrete MX_2 molecule is carbon dioxide, CO_2, and this is linear. Similarly, the only sulphide occurring as a discrete molecule, CS_2, is also linear.

Now let's see what shapes trifluorides can be. If you look at Figure 1 and your models, you will see that there are three different shapes. Boron trifluoride, BF_3, is symmetric and planar; nitrogen trifluoride, NF_3, is pyramidal; and chlorine trifluoride, ClF_3, is planar but T-shaped. These three shapes are the only arrangements of three bonds around an atom that we find in covalent molecules. Some examples of these shapes are:

Trigonal (planar): Hydrides and halides of Group III elements, for example BF_3. (Many of these form dimeric molecules, M_2X_6, rather than MX_3, for example B_2H_6.) Trioxides of Group VI elements, for example SO_3. These molecules have no lone pairs on the central atom.

Trigonal pyramidal: Hydrides and halides of Group V elements, for example NH_3 and PF_3. Trioxides of the noble gases; XeO_3 is the only example known. Here the central atom has one lone pair.

T-shaped: Trihalides of the halogens: for example BrF_3 and ICl_3, where the central halogen has two lone pairs.

In the next Section we shall look at a simple theory that can explain the different shapes of the trifluorides and enable us to predict the shapes of other molecules.

Carbon and silicon both form tetrafluorides that adopt a tetrahedral shape. You should be familiar with the tetrahedral arrangement of bonds around carbon from the Science Foundation Course.

■ You have met tetrahedral arrangements earlier in this Course. Can you remember where?

□ In Block 2, in the discussion of crystal structure.

The tetrahedral arrangement is a very important one but not the only one for tetrafluorides.

■ What other shapes of tetrafluorides can you find in Figure 1?

□ XeF_4 is square planar. The shape of SF_4 is rather odd at first sight, but becomes more understandable if you view it as a trigonal bipyramidal shape (the shape of PF_5), with one position unoccupied. Compare your models of SF_4 and PF_5.

The shapes of various MX_4 molecules are given below.

Tetrahedral: Hydrides and halides of Group IV elements, for example CH_4, $SiCl_4$, and GeF_4. Tetroxides of the noble gases, of which XeO_4 is the only example. All electrons round the central atom are used in bonds.

Square planar: Tetrahalides of the noble gases, for example XeF_4. There are two lone pairs on Xe.

SF_4-shaped: Tetrahalides of Group VI elements, for example $SeCl_4$. The Group VI elements have one lone pair.

Many coordination compounds of metal ions adopt tetrahedral (for example $[ZnCl_4]^{2-}$) or square-planar arrangements (for example $[Ni(CN)_4]^{2-}$), as do many non-metal complex ions, for example the sulphate(VI) ion SO_4^{2-} (tetrahedral).

The two possibilities for five coordination found in covalent molecules are represented by PF_5 (trigonal bipyramid) and BrF_5 (square pyramid). Note that in both PF_5 and

BrF_5 not all the bonds are the same length. The only possible arrangement of five bonds in three-dimensional space in which all five are equivalent is a planar arrangement. This would be very crowded, however, so one of the two shapes shown in Figure 1 is adopted.

However, for hexafluorides, there is an arrangement in which all six bonds are equivalent. This is the octahedron, which is illustrated by SF_6. There are not many octahedral molecules, but this arrangement is extremely important for crystal and coordination chemistry, as you saw in earlier Blocks.

Finally, we have IF_7. This is the only molecule of this type, and its shape is a pentagonal bipyramid. The arrangement of seven bonds poses the same problem as the arrangement for five bonds did, but, within experimental error, IF_7 appears to have seven identical I—F bond lengths.

You have seen that molecules of similar formula, for example NF_3 and BF_3, or PF_5 and BrF_5, can adopt different shapes. We shall now go on to study a theory that explains these different shapes and allows us to predict the shape of a molecule or ion. This theory is valence-shell electron-pair repulsion theory, and is the subject of the next Section.

4 VALENCE-SHELL ELECTRON-PAIR REPULSION THEORY

Valence-shell electron-pair repulsion (VSEPR) theory is a long name, but the principles are very simple. It is based on the fact that electrons, as negatively charged species, will tend to repel each other and therefore keep as far apart from each other as possible. In this theory we need to consider only the outer or valence-shell electrons. These electrons, as we saw when we considered the formation of covalent bonds, tend to pair up. The theory says that these *pairs of electrons will arrange themselves so as to be as far apart as possible from other pairs*. How do we apply this principle to the shapes of molecules? Below we give you some rules that will enable you to predict the arrangement of bonds around an atom and hence the shape of a molecule. We shall then go through several examples to demonstrate the use of these rules.

1 Draw a Lewis structure of the molecule, assigning valence electrons to covalent (including dative) bonds as far as possible. To do this you have to decide what sort of bond will join any two atoms. You can use the following guidelines:

(a) Hydrogen and fluorine will only form one single covalent bond to another atom.

(b) Oxygen forms two single bonds or one double bond.

(c) Chlorine, bromine and iodine will form only one single bond with elements other than oxygen and the halogens.

(d) Atoms of elements in the second row of the Periodic Table (lithium to neon) never have more than eight valence-shell electrons around them.

2 Arrange the valence-shell electrons not used for bonding into lone pairs as far as possible. You need to do this only for atoms bonded to more than one other atom.

3 Each bond and each lone pair around a particular atom is a **repulsion axis**. Note that a double or triple bond counts as only one axis. This is because the four or six electrons in the bond all join the same two atoms and therefore must all be in the same region of space.

4 Arrange the repulsion axes so that they are as far apart as possible.

Before we can apply rule 4, we need to know what arrangements of two, three, four, etc., repulsion axes ensure that they are as far apart as possible. Figure 2 shows the arrangements of two, three, four, five and six repulsion axes. These are the numbers of axes we shall come across most often. You may recognize the shapes of beryllium difluoride (BeF_2), boron trifluoride (BF_3), tetrafluoromethane (CF_4), phosphorus pentafluoride (PF_5), and sulphur hexafluoride (SF_6).

Let's see how this theory is applied to the fluorides that we studied in Section 3. We shall start with beryllium difluoride, BeF_2.

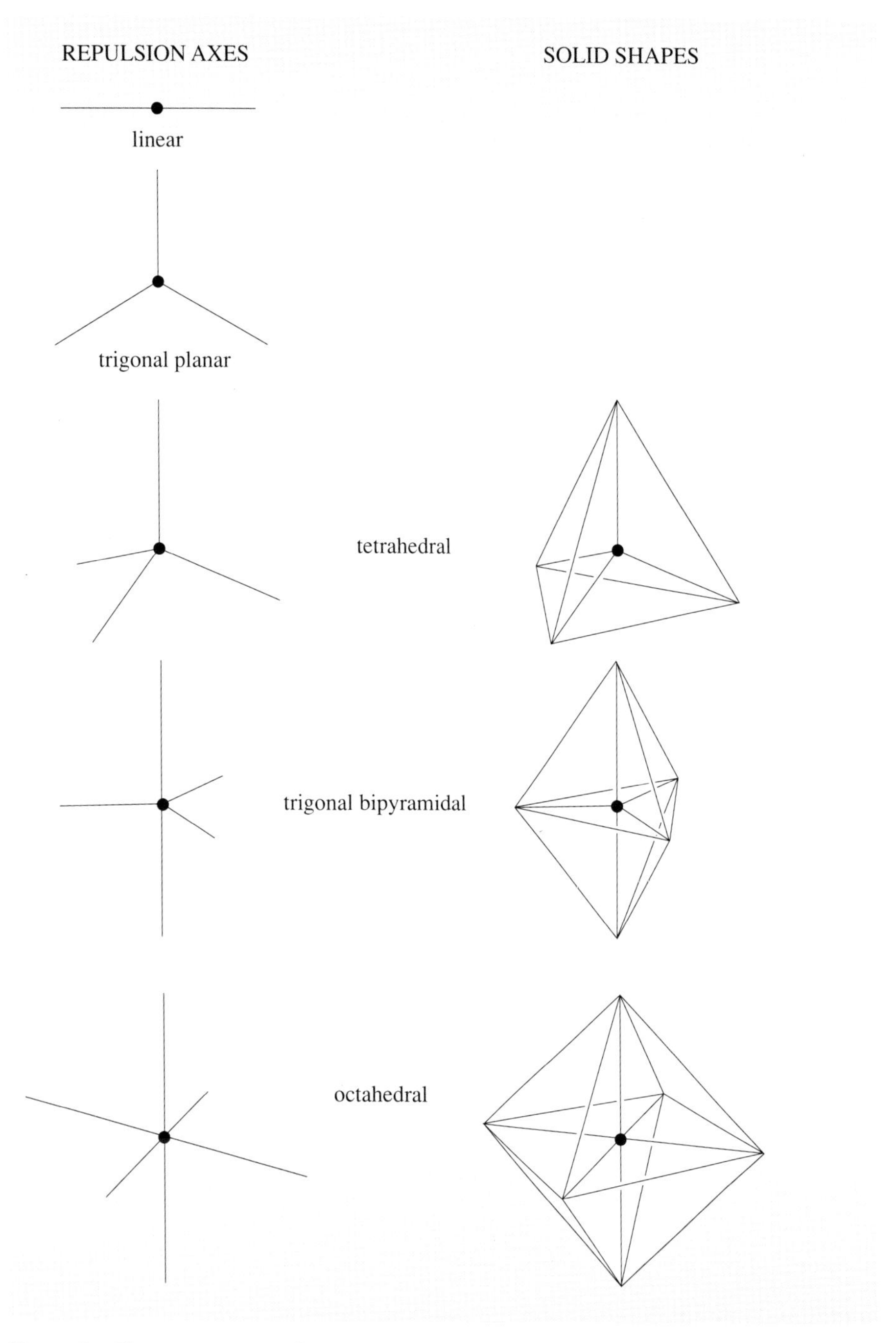

Figure 2 The arrangements of two, three, four, five and six repulsion axes. (The solid figures after which the arrangements are named are indicated by the coloured lines).

4.1 FLUORIDES WITH TWO REPULSION AXES

■ Draw a Lewis structure and structural formula for BeF_2.

□ Be has the electron configuration $1s^2 2s^2$, so it has only two valence electrons. It uses these to form two single covalent bonds to fluorine, so the Lewis structure and structural formula of BeF_2 are

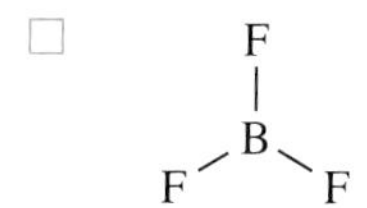

F—Be—F

The two bonds make two repulsion axes and there are no lone pairs, so we have only two axes to arrange. As you can see from Figure 2, two repulsion axes adopt a linear arrangement. VSEPR theory predicts a linear molecule and, as we saw in the last Section, BeF_2 is indeed linear. Let's now see how our theory applies to fluorides with more than two repulsion axes.

4.2 FLUORIDES WITH THREE REPULSION AXES

The shape of the BF_3 molecule is the same as the arrangement of three repulsion axes shown in Figure 2. Does the Lewis structure of BF_3 suggest that there would be three repulsion axes around the boron atom?

■ We drew the structural formula for BF_3 in Section 2. See if you can still draw it. Show any lone pairs on the boron atom.

□

There are three repulsion axes around boron. So, for BF_3, the VSEPR theory again predicts the observed shape.

4.3 FLUORIDES WITH FOUR REPULSION AXES

Four repulsion axes adopt a tetrahedral arrangement, and this is also the shape of CF_4.

■ How many repulsion axes are there around the carbon atom in CF_4?

□ Four. The four C—F bonds are repulsion axes and there are no lone pairs on the carbon atom. The structural formula of CF_4 is

```
    F
    |
F — C — F
    |
    F
```

Four repulsion axes adopt a tetrahedral shape and so CF_4 is tetrahedral.

Now suppose we have a fluoride in which the central atom is surrounded by four repulsion axes, one of which is not a bond to fluorine but a lone pair.

■ What formula would such a fluoride have, and what shape would it adopt?

□ There are only three bonds to fluorine in such a molecule so the formula would be MF_3, where M represents the central atom. The four repulsion axes would still be arranged tetrahedrally, but we would observe that the MF_3 molecule had a trigonal pyramidal shape as in Figure 3.

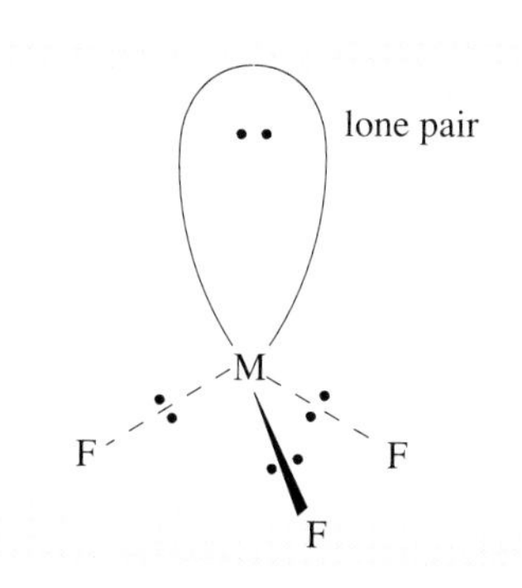

Figure 3 MF_3, showing the position of the lone pair.

■ Are any of the trifluorides in Figure 1 trigonal pyramidal?

□ Yes, NF_3.

■ Draw a Lewis structure and structural formula for NF_3.

□

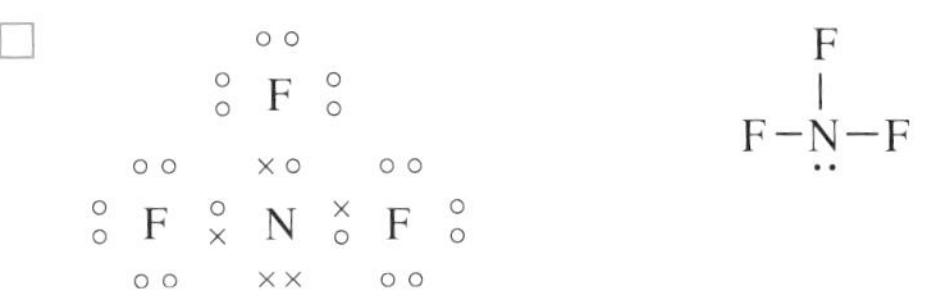

■ How many repulsion axes are there around the nitrogen?

□ There are four – three bonds and one lone pair.

The difference in shape between BF_3 and NF_3 is thus due to the lone pair on nitrogen in NF_3.

The nitrogen atom, then, will be surrounded by a tetrahedron of repulsion axes, comprising one lone pair and three fluorine atoms as shown in Figure 3.

Is this predicted shape like the observed shape in Figure 1? The shape is similar but the observed FNF bond angle is only 102.2°, considerably smaller than the regular tetrahedral angle (for example in CF_4) of 109.5°. This brings us to a refinement of our theory. In NF_3 not all four repulsion axes are equivalent, and we might expect that repulsion between a lone pair and a bond pair of electrons will not be the same as the repulsion between two bond pairs. We can explain a lot of deviations from regular shapes by assuming that the repulsions between electron pairs decrease in the following order:

lone pair : lone pair > lone pair : bond pair > bond pair : bond pair

Thus, a lone pair effectively needs more room than a bond pair. If we apply this to NF_3, we would predict a reduction in the FNF bond angle from the tetrahedral angle as the N—F bonds are squashed together to avoid the lone pair.

There are some further refinements to the theory that we shall discuss later on, and so the full set of rules and the shapes in Figure 2 are repeated on the fold-out page at the front of this Block.

The fluoride of oxygen, OF_2, also has four repulsion axes around the central atom, but two of these axes are lone pairs.

■ Assuming, as in NF_3, that the lone pairs will take up more room than the O—F bonds, what would you expect the bond angle in OF_2 to be?

□ If all four repulsion axes were equivalent, then the angle would be 109.5°. However, the lone pairs take up more room, so the O—F bonds will be pushed closer together. We would therefore expect the bond angle to be slightly less than 109.5°.

The observed bond angle is 103°, so our assumption leads to a prediction that fits the experimental results.

4.4 FLUORIDES WITH FIVE REPULSION AXES

First we shall consider phosphorus pentafluoride. In this molecule, phosphorus forms five single bonds to fluorine and these act as five repulsion axes. All the valence electrons around phosphorus are used to form these bonds, so there are no lone pairs on phosphorus. Five repulsion axes form a trigonal bipyramid (Figure 2), and this is the shape adopted by PF_5.

We mentioned earlier that the shape of sulphur tetrafluoride could be thought of as a trigonal bipyramid with one position not occupied by an atom. If this molecule obeys VSEPR theory, then we would expect the fifth position to be occupied by a lone pair. Let's see if the structural formula for SF_4 shows four bonds and one lone pair round sulphur. Sulphur has six valence electrons, so after it has used four to form bonds to fluorine it has two left over. Thus, there are four bonding pairs and one lone pair of valence electrons around the sulphur atom in SF_4, making five repulsion axes, which adopt a trigonal bipyramidal shape. Four of the five positions in the bipyramid are occupied by fluorines and the fifth by a lone pair of electrons.

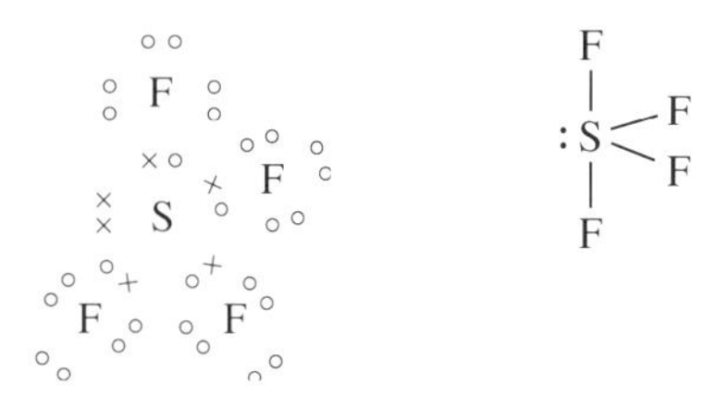

We have one problem left with SF_4 though. In a trigonal bipyramid not all positions are equivalent. Look at your model of PF_5. We can distinguish two different sorts of position (Figure 4). One is that occupied by the three fluorines that form a planar triangle with phosphorus in the middle. These are called **equatorial** atoms. The other sort of position is called **axial**.

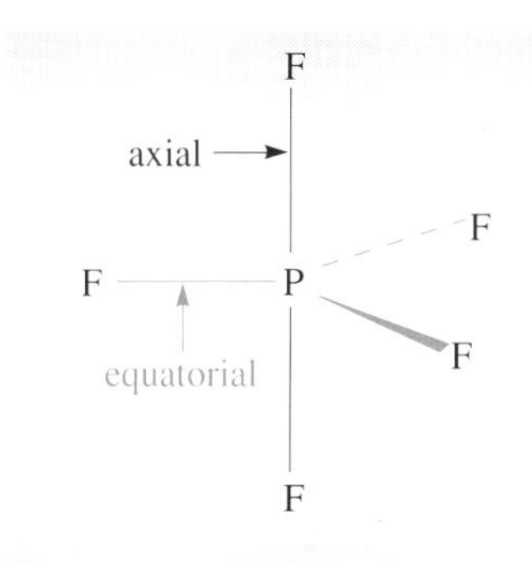

Figure 4 Axial and equatorial atoms in PF_5.

Now the question is, will the lone pair in SF_4 occupy an equatorial or an axial site? Because lone pair : bond pair repulsions exceed bond pair : bond pair repulsions, we would expect the lone pair to occupy the site where lone pair : bond pair repulsions are minimized. It turns out that this site is an equatorial one, because lone pair : bond pair repulsions between axes 120° apart are much less than those between axes 90° apart. If the lone pair occupies an equatorial site, then there are two axes at 90° to it. One in an axial position would have three axes at 90° to it.

Now let's consider a fluoride in which the central atom is surrounded by five repulsion axes, two of which are lone pairs. The lone pairs occupy equatorial sites in a trigonal bipyramid. The lone pair : lone pair repulsion is reduced because the lone pairs are at 120° to each other rather than 90°, and the lone pair : bond pair repulsions are reduced as in SF_4.

■ What shape molecule would this give us?

□ Only one equatorial position and the two axial positions would be occupied by bonds to fluorine. This would give us a T-shaped trifluoride.

Chlorine trifluoride is just such a T-shaped molecule, and if you draw the structural formula of ClF_3 you will find that there are three Cl—F bonds and two lone pairs around the chlorine atom.

If all three equatorial sites in a trigonal bipyramid are occupied by lone pairs, then we will have a linear molecule. An example of a linear molecule with five repulsion axes around the central atom is krypton difluoride, KrF_2. Note that both BeF_2 and KrF_2 are linear, although there are five pairs of electrons around the krypton atom and only two pairs around the beryllium atom.

4.5 FLUORIDES WITH SIX REPULSION AXES

The fluorides with six repulsion axes around the central atom are SF_6, BrF_5 and XeF_4. In SF_6 all the valence electrons around sulphur are used to form bonds to fluorine. The six S—F bonds are thus the only repulsion axes and the molecule is octahedral.

Bromine has seven valence electrons, one more than sulphur. In BrF_5 it uses five of these to form bonds to fluorine and the remaining two electrons form a lone pair. The square-based pyramid shape of BrF_5 is thus an octahedron with one position occupied by a lone pair. All positions in an octahedron are equivalent, so it does not matter which one we choose to put the lone pair in.

■ How many repulsion axes are there around xenon in XeF_4?

□ The structural formula of XeF_4 is shown in the margin. There are six repulsion axes: four Xe—F bonds and two lone pairs.

■ These six axes will adopt an octahedral arrangement. Which positions would you expect the lone pairs to occupy?

□ The lone pairs will tend to avoid each other, and so will adopt positions on opposite sides of the octahedron.

The resulting shape is square planar, which is what we observe for XeF_4.

4.6 MOLECULES OTHER THAN FLUORIDES

We are now going on to look at other types of molecule. First of all we shall consider oxides, and we are going to start with sulphur dioxide, SO_2.

■ Draw the Lewis structure for SO_2.

□ O $\overset{\circ}{\underset{\circ}{\times}}$ S $\overset{\circ}{\underset{\circ}{\times}}$ O (with non-bonding electron pairs: two pairs on each O, one pair ×× on S)

■ Write down the structural formula for SO_2.

□ $O{=}\ddot{S}{=}O$

■ How many lone pairs are there around sulphur in SO_2?

□ One.

■ How many repulsion axes are there around sulphur in this molecule?

□ Three. Two double bonds and one lone pair. Remember that a double bond constitutes one repulsion axis.

Three repulsion axes form a trigonal planar arrangement, with angles between them of 120°. For SO_2 one repulsion axis is a lone pair so we would expect to find a bent molecule with an OSO bond angle slightly less than 120°. The measured angle is 119.5°, which is very close to 120°. This is perhaps closer than you would expect after seeing the deviations from 109.5° in NF_3 and OF_2, but remember that double bonds, as in SO_2, will take up more room than single bonds, as in OF_2. Ozone, O_3, is also a V-shaped molecule but the central oxygen cannot expand its octet and so ozone does not contain two double bonds. The Lewis structure and structural formula of O_3 are:

O $\overset{\circ}{\underset{\circ}{\times}}$ O $\times$ O (Lewis structure) $O{=}O{\rightarrow}O$

Ozone has a bond angle of 117°.

Although we have said that atoms are bonded together by pairs of electrons, there are a few examples of stable molecules with odd numbers of electrons. In this case, one of the repulsion axes is an odd electron. One unpaired electron takes up much less space than a lone pair, and in fact needs less room than a single bond.

An example of such a molecule is ClO_2; its Lewis structure and structural formula are:

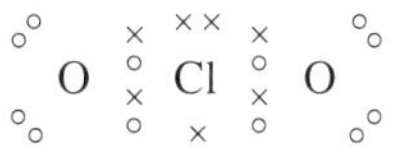

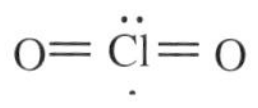

O ⁘ Cl ⁘ O $O{=}\ddot{Cl}{=}O$

Thus ClO_2 has four repulsion axes, but the effect of the odd electron is to make the OClO bond angle much greater than the tetrahedral angle (117.5°). The deviation is sufficiently large that the angle is closer to 120° than to the tetrahedral angle. An alternative way of viewing ClO_2 is to regard the odd electron as having no steric role and therefore not counting it as a repulsion axis. This would give a prediction of an angle close to 120° since there would then be three repulsion axes – one lone pair and two double bonds. There are very few **odd-electron molecules**, but if you do meet one you may either not count the single electron as a repulsion axis or take it as a repulsion axis that occupies much less space than any other.

So far we have considered only molecules (fluorides and oxides) in which one atom is attached to several identical atoms. Let's now consider some examples in which not all the surrounding atoms are the same.

■ Draw a Lewis structure and structural formula for $POCl_3$.

□

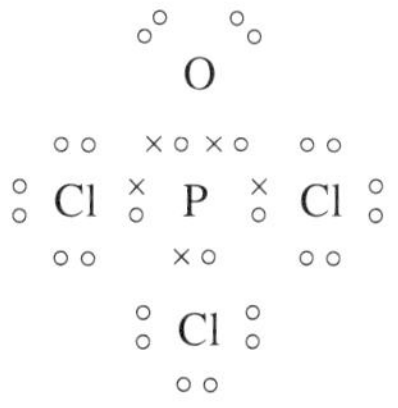

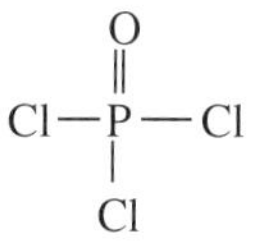

```
         O
                               O
                               ‖
 Cl  P  Cl                 Cl—P—Cl
                               |
         Cl                    Cl
```

■ How many repulsion axes are there?

□ Four. One double bond and three single bonds. All the valence electrons on the phosphorus atom are used in bonding and so there are no lone pairs.

■ What shape do four repulsion axes adopt?

□ Tetrahedral.

■ Will $POCl_3$ be a regular tetrahedron?

□ We have said nothing so far about the effect on bond angles of different atoms, but we might well expect that a P=O double bond will not be equivalent to a P—Cl single bond. A double bond, having more electrons, will repel a single bond more strongly than two single bonds repel each other. In this case we would expect the OPCl bond angles to be greater than 109.5° and the ClPCl bond angles to be less than 109.5°.

The measured ClPCl bond angle is 103.3° so it seems that our reasoning was correct.

■ Draw a Lewis structure and a structural formula for NSF (S is the central atom).

□

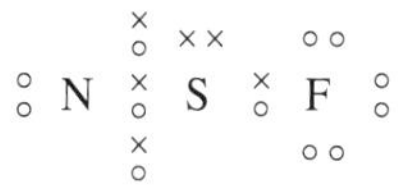

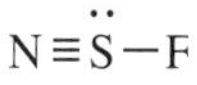

N ⁝ S ⁚ F $N{\equiv}\ddot{S}{-}F$

■ How many repulsion axes are there?

□ Three: the N≡S triple bond, the S—F single bond and the lone pair.

■ What shape do three axes adopt?

□ Trigonal planar.

■ Would you expect the NSF bond angle to be 120°?

□ Probably not. The triple bond : lone pair repulsion would be the largest so the NSF bond angle will probably be less than 120°.

The measured bond angle is 117°.

Next, we shall look at two ions, to show that VSEPR theory is not confined to neutral molecules. First consider the anion BrF_4^-, the tetrafluorobromate(III) ion.

■ The anion BrF_4^- has a central bromine atom, surrounded by four fluorines. Draw a structural formula for the ion.

□ Putting the extra electron on bromine, you should have got

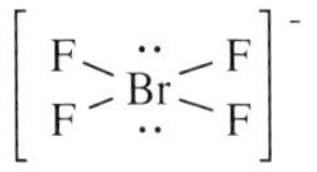

$$\left[\begin{matrix} F & & F \\ & \ddot{\underset{..}{Br}} & \\ F & & F \end{matrix} \right]^-$$

■ How many repulsion axes are there around bromine?

□ Six. Four Br—F bonds and two lone pairs.

■ What shape do six axes adopt?

□ Octahedral.

■ What positions in the octahedron will the lone pairs occupy?

□ To minimize repulsions they will occupy positions opposite each other, leaving the four fluorines in a plane with the bromine.

The square-planar arrangement is in fact the one observed for BrF_4^-. This is the same shape as XeF_4, which has the same number of valence electrons.

Crystalline phosphorus pentachloride consists of PCl_4^+ ions and PCl_6^- ions. Let us try to predict the geometry of the PCl_4^+ ion.

■ Write down a Lewis structure and a structural formula for PCl_4^+.

□

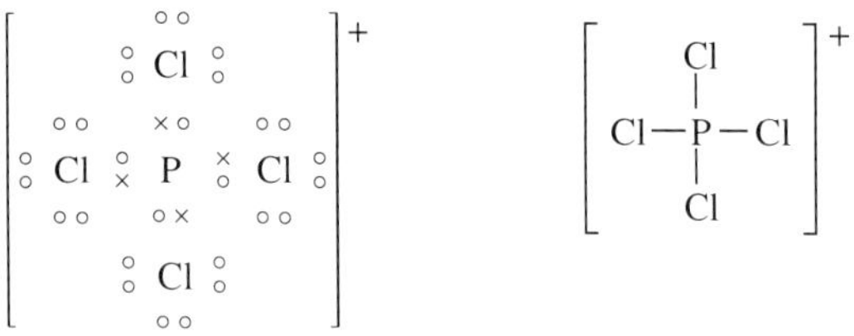

$$\left[\begin{matrix} & \overset{\circ\circ}{\underset{\circ}{\overset{\circ}{}}}Cl\overset{\circ}{\underset{\circ}{}} & \\ \overset{\circ\circ}{\underset{\circ\circ}{\overset{\circ}{\underset{\circ}{}}Cl}} \overset{\circ}{\underset{\times}{}} & \overset{\times\circ}{\underset{\circ\times}{P}} & \overset{\times}{\underset{\circ}{}} \overset{\circ\circ}{\underset{\circ\circ}{Cl}}\overset{\circ}{\underset{\circ}{}} \\ & \overset{\circ}{\underset{\circ}{}}\underset{\circ\circ}{Cl}\overset{\circ}{\underset{\circ}{}} & \end{matrix} \right]^+ \qquad \left[\begin{matrix} & Cl & \\ & | & \\ Cl- & P & -Cl \\ & | & \\ & Cl & \end{matrix} \right]^+$$

■ How many repulsion axes are there round phosphorus?

□ Four.

So the ion should be tetrahedral. The experimentally determined structure shows that this prediction is correct.

Well, our theory seems to be successful so far, and so we shall finish this Section by looking at some molecules for which it doesn't work so well. Our first example is barium difluoride. The BaF_2 molecule, like BeF_2, can be obtained in the vapour phase by heating the solid.

Like beryllium, barium has two valence electrons, which it uses to form two single bonds to fluorine. There are therefore only two repulsion axes and we would expect a linear molecule. The BaF_2 molecule, however, is bent, with an estimated bond angle of 100°, a lot smaller than the expected 180°. Other examples where the theory does not work are the ions $[InCl_5]^{2-}$, which is a square pyramid rather than the expected trigonal bipyramid, and $[TeCl_6]^{2-}$, which is a regular octahedron despite having six bonds *and* a lone pair round the tellurium atom. (The shapes of these two ions are affected by surrounding ions in the crystal lattice.) If you go on to study chemistry at third level, you will find many exceptions to these rules in the compounds of the transition metals.

However, the valence-shell electron-pair repulsion theory provides a simple way of predicting the rough shape of a molecule, and is a good starting point for more accurate calculations and for accurate experimental determinations of bond lengths and angles.

SAQ 6 Use valence-shell electron-pair repulsion theory to predict the shapes of the following molecules: (a) hypochlorous acid, HOCl; (b) boron hydrogen difluoride, HBF_2 (B central atom); (c) carbon dioxide, CO_2 (OCO); (d) hydrogen cyanide, HCN; (e) sulphuryl fluoride, SO_2F_2, which has its atoms joined in the following order (this is not a structural formula):

```
O   F
 \ /
  S
 / \
O   F
```

SAQ 7 Use valence-shell electron-pair repulsion theory to predict the shapes of the following ions: (a) ICl_2^- (central atom I); (b) PCl_6^- (central atom P); (c) the nitrate anion, NO_3^- (central atom N); (d) NO_2^+ (ONO^+).

SAQ 8 Would you expect the approximate shapes of the molecules and ions below, all of which contain five atoms, to be (a) tetrahedral, (b) square planar (as XeF_4) or (c) like SF_4?

(i) AlH_4^-, (ii) BiI_4^-, (iii) CCl_4, (iv) GeI_4, (v) NH_4^+, (vi) PSF_3 (central atom P), (vii) $SeCl_4$.

SAQ 9 The experimentally determined shapes of some molecules are given below. In each case, state whether you think valence-shell electron-pair repulsion theory explains the shape.

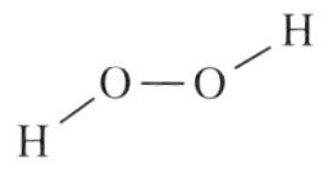

H–O–O–H

(a)

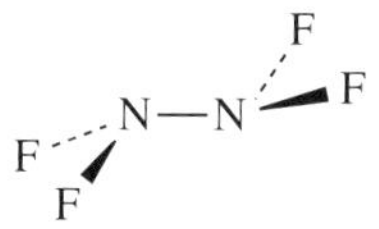

F_2N–NF_2

(b)

O_3S–NH_2

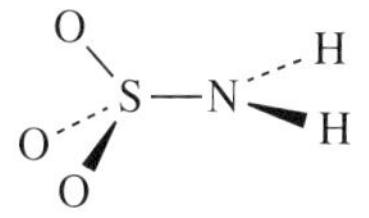

(c)

5 SYMMETRY

Chemists have adopted from mathematicians a way of classifying molecules according to their shapes. The classification is based on the idea that if a molecule is sufficiently symmetric, then an action such as turning the molecule half way round will leave the molecule looking the same as it did when you started. Actions such as rotating a molecule are called **symmetry operations**, and a molecule can be placed in a category according to how many such operations you can perform and still leave the molecule looking the same.

Such a classification has proved particularly useful in calculating chemical properties and in predicting and analysing molecular spectra. For example, if I wish to calculate a property of the molecule SF_6, I need to tell the computer only the length of one S—F bond and the shape. The computer then knows that all the S—F bonds are similar and can save time and computer space by doing sums only once instead of six times. In spectroscopy, the classification will allow you to decide relatively easily, for example, whether a transition between two particular energy levels will occur and hence give rise to a line in the spectrum. You will see in a later Block how symmetry is applied to spectroscopy. Here we are concerned with introducing the classification scheme and applying it to covalent bonding.

STUDY COMMENT Video sequence 5, entitled *Symmetry*, is designed to reinforce your study of Section 5. You should watch it before or during your reading of this Section.

5.1 ROTATION

The objects in Figure 5 have something in common. If asked what it was, you might suggest it has something to do with three – the objects have three points or three spokes or three sides etc. This is correct, but in order to use the symmetry classification, we have to find an *action* that will leave all the objects looking the same. Now, if they are going to look the same, each spoke, point, line, etc., must move to the position of the next one. One thing we could do is turn each object round so that each corner of the triangle, for example, has moved round to the position of the next one. If the corners of the triangle are labelled a, b and c then the result of this action is shown

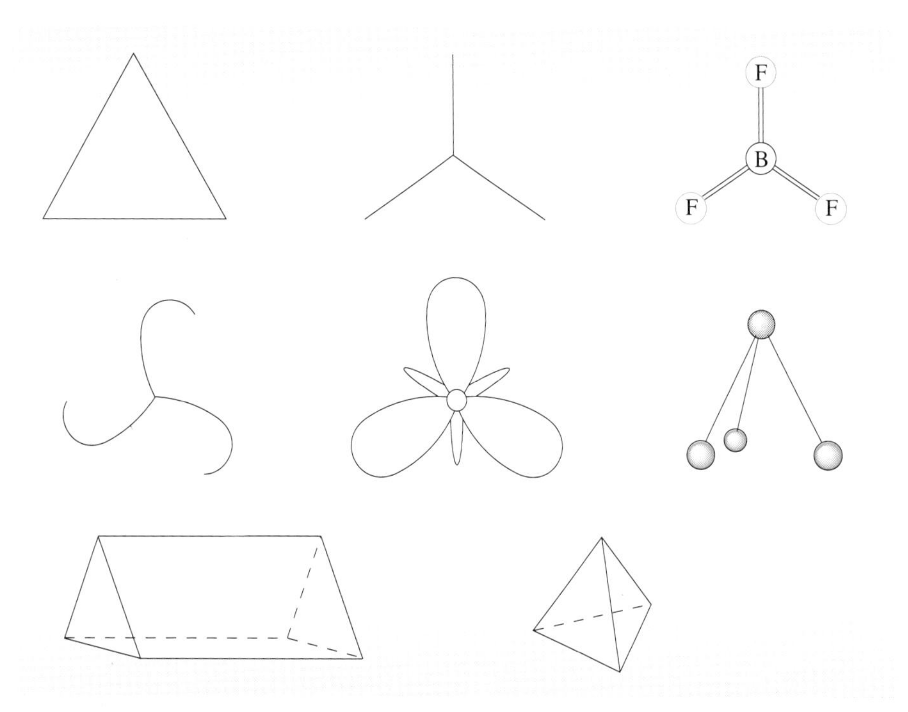

Figure 5 A set of objects.

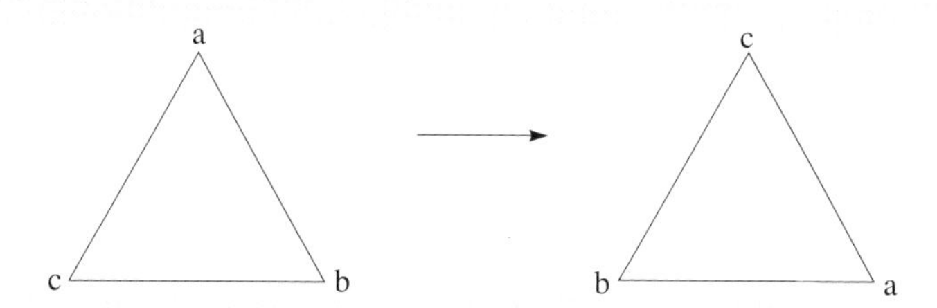

Figure 6 The result of rotating an equilateral triangle through a third of a revolution.

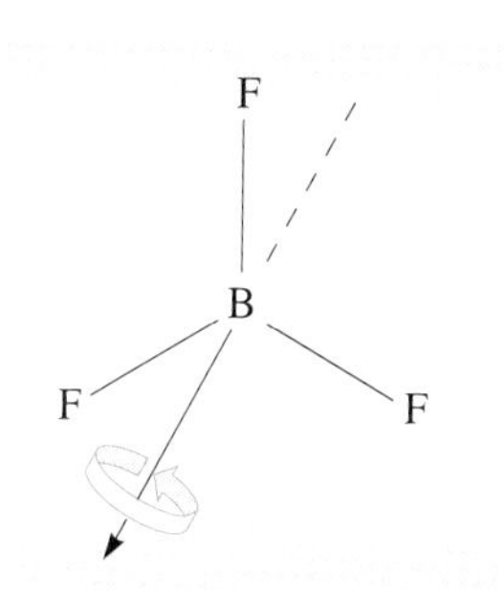

Figure 7 A threefold axis in BF_3.

in Figure 6. The same action will produce an object identical with the one you started with when performed on all the objects in Figure 5. So the action is a symmetry operation. What we have done is rotate the object through one third of a complete revolution. Such an action is a rotation, and because three such rotations return the object to its starting position, it is known as a threefold rotation.

For a complete description of the symmetry operation, we need to specify what the objects are rotating about. If you consider a wheel on a cart or a buggy, the wheel rotates about an axle. We can imagine a sort of axle going through the objects in Figure 5. This is illustrated for the BF_3 molecule in Figure 7. The line about which the molecule is rotated is called an **axis of symmetry,** in this case a threefold axis of symmetry. An axis of symmetry is an example of a **symmetry element**.

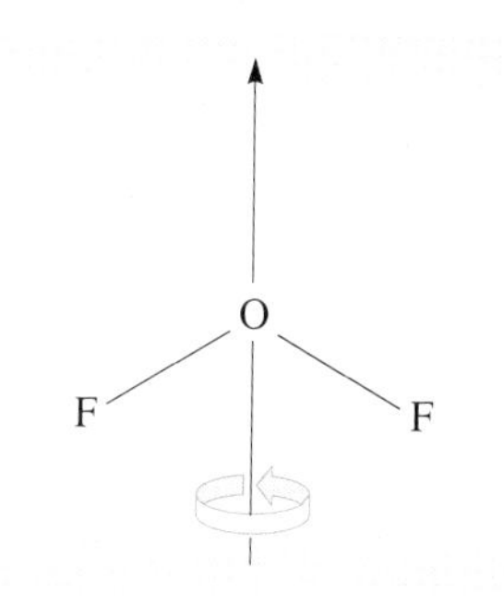

Figure 8 The twofold axis of symmetry in OF_2.

The objects in Figure 5 had a threefold axis of symmetry, but other objects need to be turned through different amounts to remain identical. OF_2, for example, has a twofold axis of symmetry (Figure 8). Rotation through half a complete revolution about the line shown produces an identical-looking molecule.

Both XeF_4 and SF_6 have fourfold axes. A quarter turn produces an identical-looking molecule, and four such turns are needed to take the molecule back to its starting point. The fourfold axis of XeF_4 and one of the fourfold axes of SF_6 are shown in Figure 9.

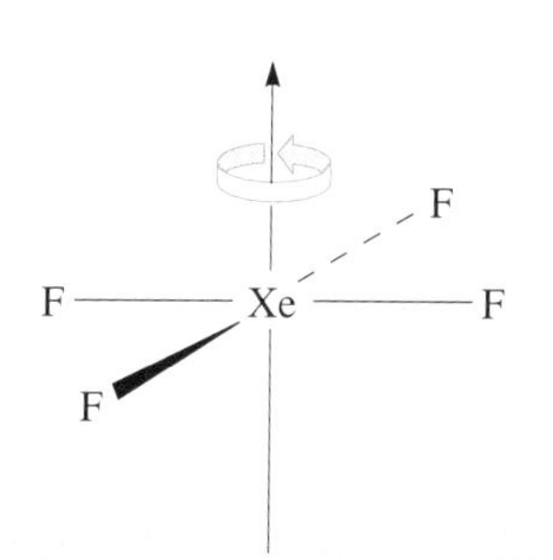

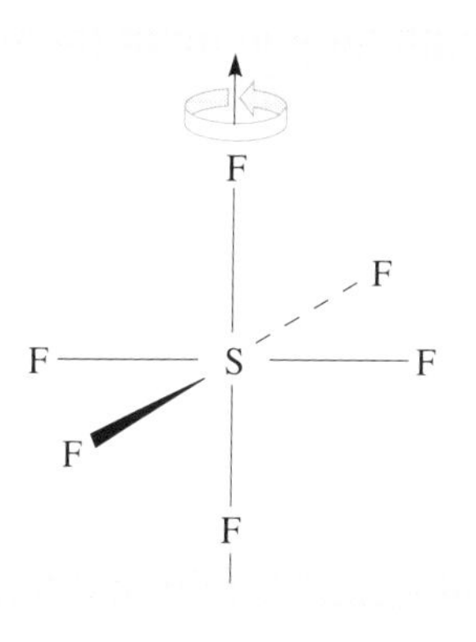

Figure 9 Fourfold axes in XeF_4 and SF_6.

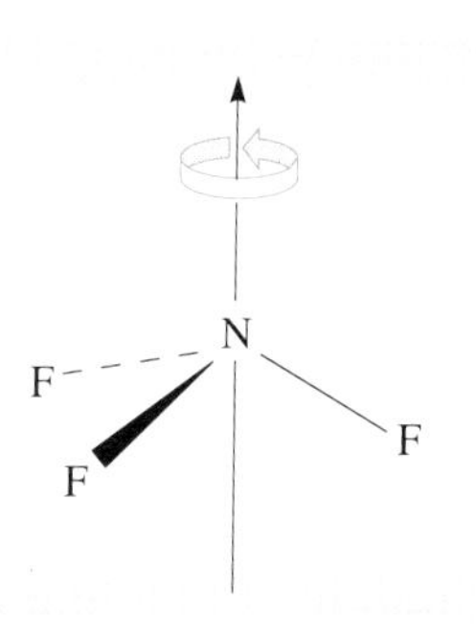

Figure 10 The threefold axis in NF_3.

■ Has NF_3 an axis of symmetry?

□ Yes: it has a threefold axis of symmetry, which goes through the nitrogen atom. This is shown in Figure 10. You may find it helpful to build a model of NF_3 and rotate it through a third of a revolution.

Both BF_3 and NF_3 have threefold axes but they are not the same shape. BF_3 is trigonal planar and NF_3 is pyramidal. Our classification scheme allows us to distinguish these two shapes by considering what other symmetry operations might apply to the two. The first thing to note is that molecules (or other objects) may have more than one axis of rotation.

Suppose we rotate BF_3 about a B—F bond. After half a turn we arrive at something that looks identical with the starting molecule. Two such turns are required to restore us to the starting point and so the B—F bond is a twofold axis (Figure 11).

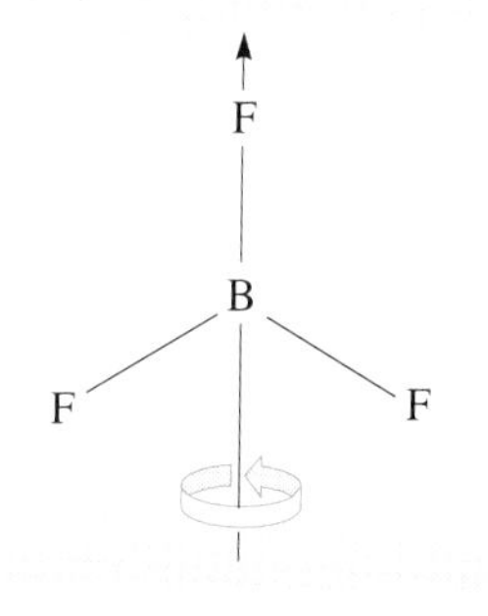

Figure 11 A twofold axis in BF_3.

There are three B—F bonds and no reason for choosing one rather than another, so each B—F bond is a twofold axis. BF_3 therefore has three twofold axes.

A situation like this in which a molecule has an n-fold axis of symmetry and n twofold axes at right-angles is not uncommon, and so it is worth looking for n twofold axes when you have found an n-fold axis (n is 2 or higher).

Rotating NF_3 about an N—F bond, by contrast, produces the result shown in Figure 12.

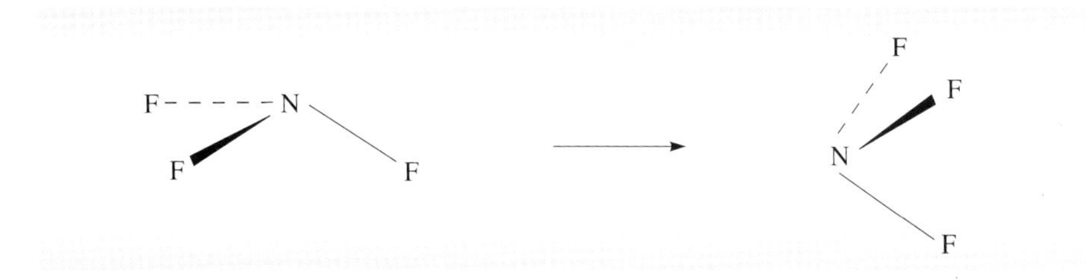

Figure 12 Rotation of NF_3 by half a revolution about an N—F bond.

The N—F bonds in NF_3 are *not* axes of symmetry. NF_3 in fact has only one axis of symmetry, the threefold axis already discussed. The shapes of NF_3 and BF_3 can thus be distinguished by the number of symmetry axes.

■ How many symmetry axes has XeF_4?

□ Earlier we said that XeF_4 has a fourfold axis. We look to see if there are any others. A good starting place is to see if there are n twofold axes at right-angles to the n-fold axis. In this case $n = 4$ and so we look for four twofold axes. The fourfold axis is at right-angles to the plane of the molecule (Figure 9) and so these twofold axes, if they exist, must be in the plane. An axis running through two opposite Xe—F bonds is a twofold axis (Figure 13) and there are two such axes. There are another two axes in this plane, lying between Xe—F bonds (Figure 14). So XeF_4 has one fourfold axis and four twofold axes.

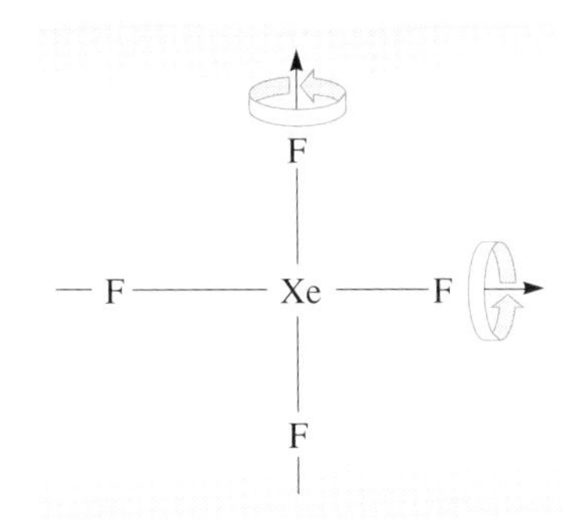

Figure 13 Twofold axes along Xe—F bonds in XeF_4.

There cannot be an axis of higher order than fourfold in XeF_4 because there are only four fluorine atoms and, at each turn, an atom must move to a position occupied by an atom identical with itself. Is there a threefold axis? Well, a threefold axis would interchange three fluorines and so such an axis would have to lie along an Xe—F bond. But the Xe—F bonds are only twofold axes and so there is no threefold axis.

One fourfold axis and four twofold axes are thus the total for XeF_4.

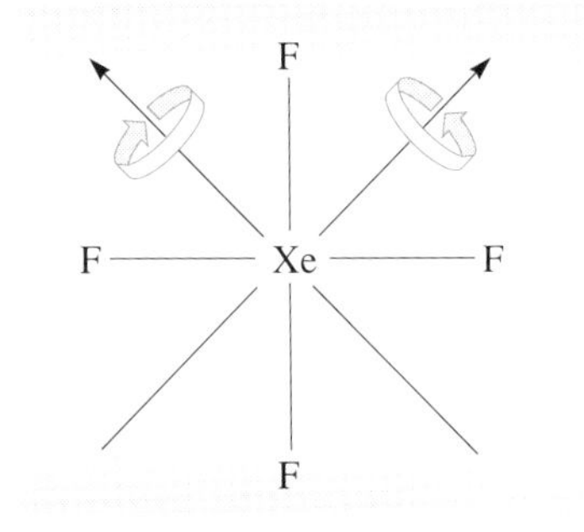

Figure 14 Twofold axes between Xe—F bonds in XeF_4.

Now let's see how many axes we can find for SF_4. If you look carefully at the diagram of SF_4 in Figure 1, you will see that the four fluorine atoms do not all occupy similar positions. For example, two S—F bonds have an angle between them of 103° but none of the other possible pairs of bonds have this angle between them. We can divide the fluorines into two sets of two. The top and bottom pair, called axial, form one set, and the two separated by an FSF angle of 103°, known as equatorial, form the other set. No symmetry operation will move a fluorine from one set to a position occupied by a fluorine in the other set. Such an operation would leave the molecule looking different from the way it started. The four fluorine atoms in XeF_4 can all be interchanged by symmetry operations because they all occupy similar sites, but in SF_4 the axial and equatorial sites are distinct.

Since we cannot move an axial fluorine to an equatorial position by a symmetry operation, you need look only for operations that move axial fluorines to axial positions and equatorial fluorines to equatorial positions.

There are only two fluorine atoms in each set and so the highest order axis SF_4 could possibly have is a twofold axis.

■ Does SF_4 have a twofold axis of symmetry?

□ Yes. If we imagine a line in the plane of the equatorial fluorines and lying between the two equatorial S—F bonds as in Figure 15, then rotation through half a revolution about this axis will swap the axial fluorines with each other and the equatorial fluorines with each other. The result will be an identical-looking molecule and so this is a twofold axis. This is the only axis of symmetry for SF_4.

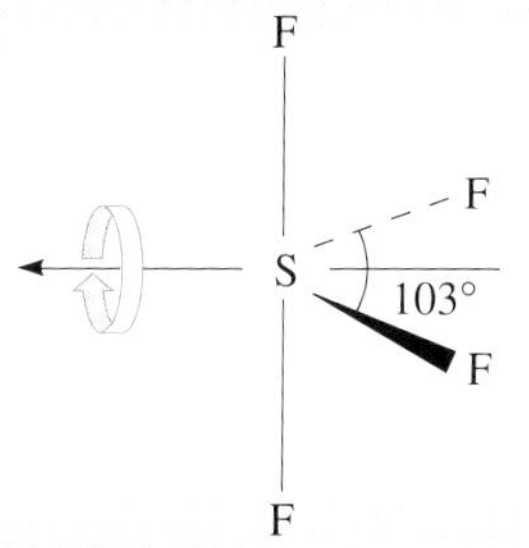

Figure 15 The twofold axis of SF_4.

So, as a general strategy for molecules like the fluorides, with a central atom surrounded by n atoms of the same type, you first of all divide the n atoms into sets occupying similar positions. If the largest number of atoms in a set is m, then the highest order axis you need look for is m-fold. You may not find an m-fold axis; there are some molecules (for example CF_4 and SF_6) for which, although all n atoms are in one set, there is no n-fold axis. This is because the atoms cannot be interchanged by rotation unless they lie in a plane. If you cannot find an m-fold axis, look for a lower order one ($m - 1$, $m - 2$, etc.).

When you have found an axis of order 2 or higher, look for n twofold axes at right-angles to this axis.

There is one special case that we must consider and that is linear molecules. If we take BeF_2, there are only two fluorines so normally we would look only for a twofold axis. But in the case of linear molecules, there is a symmetry axis running along the molecular axis. Now, if we rotate the molecule about this axis, it remains looking the same however much we turn it. Because we could rotate it by an infinitesimal amount, we call this axis an infinite axis of symmetry. All linear molecules have such an axis. As with non-linear molecules we still have to look for n twofold axes at right-angles to the infinite axis, and in this case it means looking for an infinite number of such axes. In BeF_2 there are an infinite number of twofold axes, and Figure 16a illustrates one of them. We have drawn this axis in the plane of the paper, but any axis at right-angles to the molecular axis at any angle to the plane of the paper will be a twofold axis as well. Figure 16b shows the molecule viewed along its infinite axis, and a few of the infinite number of twofold axes.

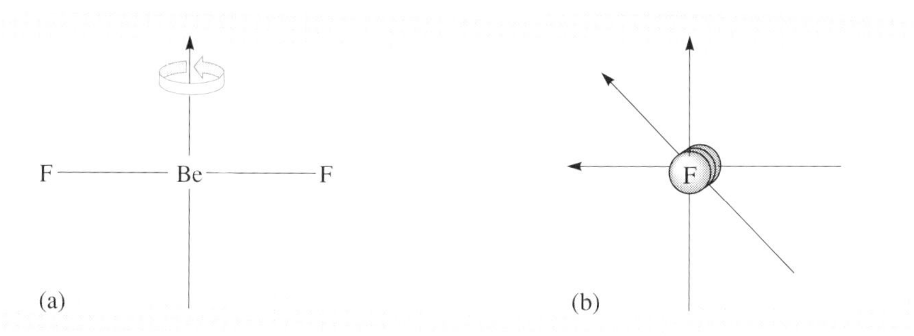

Figure 16 Twofold axes in BeF_2.

■ What are the axes of symmetry in a less symmetrical linear molecule, such as HCN?

□ HCN will have an infinite axis because it is linear. It does not, however, have any twofold axes because there are no identical atoms to interchange.

Now try the following SAQs to practise finding axes of symmetry, and then go on to see what other symmetry operations there are.

SAQ 10 Find all the axes of symmetry in the following molecules, and state whether they are twofold, threefold, etc.: (a) BrF_5; (b) SO_2; (c) $POCl_3$; (d) NO_3^-; (e) HBF_2.

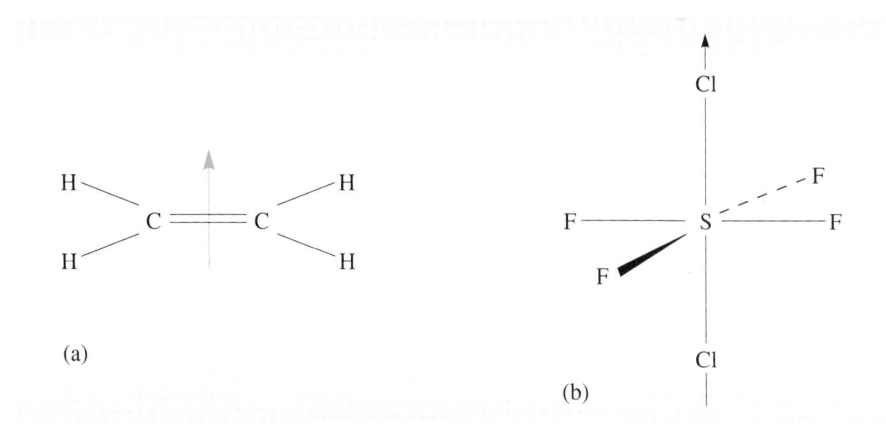

Figure 17 (a) An axis in ethene. (b) An axis in SCl_2F_4.

SAQ 11 Are the axes (a)–(d) axes of symmetry? If so, what is their order (twofold, threefold...)?

(a) A line along one C—F bond in CF_4.

(b) A line through the opposite S—F bonds in SF_6.

(c) A line at right-angles to the C—C bond and going through the centre of this bond in ethene (Figure 17a).

(d) A line through the two S—Cl bonds in SCl_2F_4, with the two chlorines opposite (Figure 17b).

5.2 REFLECTION

The molecule NFH_2

N
F H
H

looks more symmetrical than NFHCl

N
F Cl
H

but neither molecule has an axis of symmetry. Is there a symmetry operation that will distinguish these two?

Yes, it is called reflection. If we draw a plane that contains the N—F bond and lies between the two N—H bonds as in Figure 18, then a line drawn from one hydrogen atom to the plane and out the other side by the same amount meets the other hydrogen atom. The reflection of each hydrogen in the plane is the other hydrogen. This is our second symmetry operation – reflection through a **plane of symmetry**.

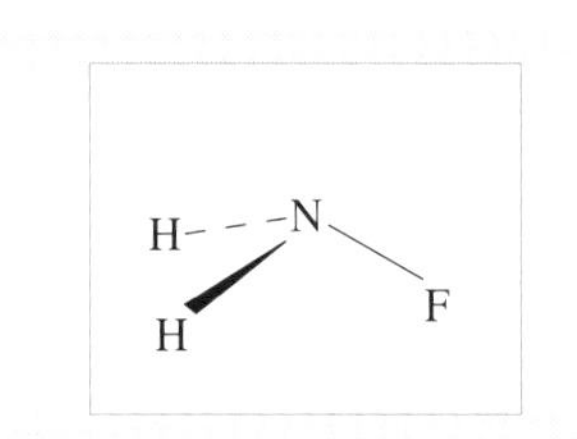

Figure 18 A symmetry plane in NFH_2.

NFHCl has no plane of symmetry, and so this distinguishes it from NFH_2.

NFH_2 has only a plane of symmetry, but many molecules have both plane(s) and an axis (or axes) of symmetry. NF_3, for example, has a plane of symmetry just like that in Figure 18, but with fluorine atoms rather than hydrogen atoms being reflected. Because all three atoms bonded to nitrogen are the same in NF_3, however, it does not matter which N—F bond lies in the plane and so there are three such planes (Figure 19).

NF_3 has a threefold axis of symmetry as well as the three planes of symmetry. The three planes each contain this axis. NF_3 has only one axis of symmetry and so this threefold axis is the **principal axis** of NF_3, the axis of highest order. Planes of symmetry containing the principal axis are known as **vertical planes**. As it is usual to denote the principal axis of the molecule as the z axis and to have the z axis pointing up the page, then such planes of symmetry would indeed be vertical when the molecule is oriented in the standard manner. NF_3, then, has one threefold axis of symmetry and three vertical planes of symmetry. Molecules with an n-fold principal axis and n vertical planes of symmetry are common, and so it is worth looking for these planes when you have found the principal axis.

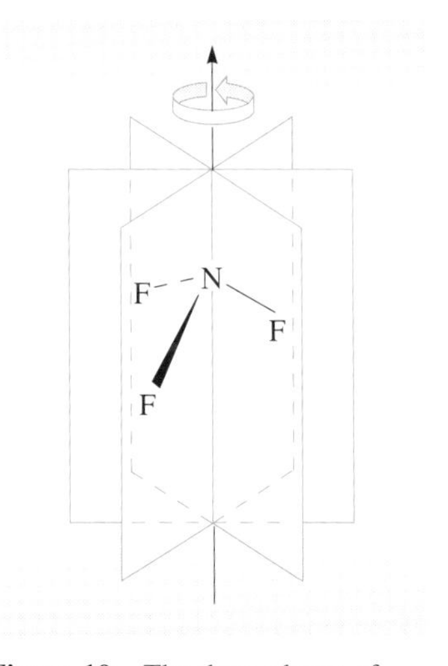

Figure 19 The three planes of symmetry in NF_3.

Now consider BF_3. This molecule, like NF_3, has as its principal axis a threefold axis. Does it have three vertical planes of symmetry? Yes: as in NF_3, each fluorine lies in a plane of symmetry that reflects the other two fluorines into each other. We saw that BF_3 had more axes of symmetry than NF_3; has it more planes of symmetry than NF_3?

BF_3 is a planar molecule. What happens if we reflect each atom through this plane? The molecule is unchanged, because all the atoms lie in the plane and so do not move at all when reflected. The plane of the molecule is thus a plane of symmetry. However, this is not a vertical plane of symmetry. The principal axis of BF_3 is at right-angles to the plane; if we arrange the molecule with the principal axis vertical then the molecular plane is horizontal. A plane of symmetry at right-angles to the principal axis is known as a **horizontal plane**; BF_3 thus has three vertical planes of symmetry and one horizontal plane of symmetry.

It is useful to remember that all planar molecules must have a plane of symmetry – the plane of the molecule; however, this is not necessarily horizontal, as you will see in the next example.

■ Has OF_2 any planes of symmetry?

□ Yes. OF_2 is planar and thus the molecular plane is a plane of symmetry. However, the principal axis of OF_2 (the twofold axis – its only axis of symmetry) lies in the molecular plane. The molecular plane of OF_2 is thus a *vertical* plane of symmetry. OF_2 also has another plane of symmetry. This lies between the two O—F bonds and is also vertical. The two planes are shown in Figure 20.

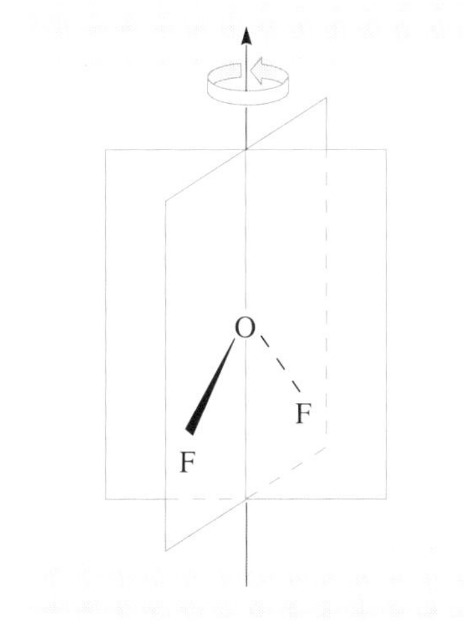

Figure 20 Planes of symmetry in OF_2.

The plane of symmetry in NFH_2 is neither vertical nor horizontal because NFH_2 has no axis of symmetry. It is referred to simply as a plane of symmetry.

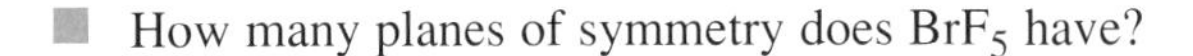

■ How many planes of symmetry does BrF_5 have?

□ BrF_5 has four planes of symmetry (Figure 21). All contain the Br—F bond that lies on the fourfold axis. Two of them contain two other Br—F bonds. The other two lie between Br—F bonds. Since they all contain the principal (fourfold) axis, the planes are all vertical.

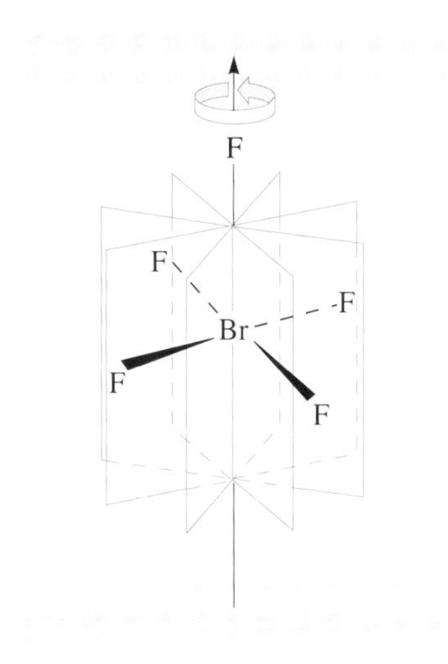

Figure 21 Planes of symmetry in BrF_5.

Note that BrF_5, like OF_2 and NF_3, has n vertical planes where the principal axis is n-fold; n in this case is 4.

SAQ 12 How many planes of symmetry are there in (a) PF_5, (b) SF_4 and (c) IF_7?

SAQ 13 The descriptions (i) to (vii) are of planes in molecules. For each one decide whether it is (a) a vertical plane of symmetry, (b) a horizontal plane of symmetry, (c) a plane of symmetry that is neither vertical nor horizontal, or (d) not a plane of symmetry at all.

(i) The molecular plane of HOF:

```
   O
  / \
 H   F
```

(ii) The molecular plane of *trans*-1,2-difluoroethene:

```
 H     F
  \   /
   C=C
  /   \
 F     H
```

(iii) A plane containing the P=O bond and one P—Cl bond in $POCl_3$.

(iv) A plane through the carbon atom and at right-angles to the molecular axis in CO_2.

(v) A plane containing the HF molecule.

(vi) The plane of the three fluorine atoms in NF_3.

(vii) The molecular plane of NO_3^-.

5.3 INVERSION

The final symmetry operation that you will be expected to recognize in this Course is inversion through a **centre of symmetry**. To invert through a centre of symmetry, we imagine a line drawn from any atom in the molecule to the centre of the molecule. We then continue this line until it is the same length the other side. If the molecule has a centre of symmetry then the line will end up on an identical atom regardless of which atom we started with. This molecule, for example, has a centre of symmetry in the middle of the benzene ring:

```
      H     Cl
       \   /
        C-C
       /   \
  Br-C       C-Br
       \   /
        C-C
       /   \
     Cl     H
```

A line from a bromine to the centre and out the other side brings us to a bromine, from a chlorine to a chlorine, a carbon to a carbon and a hydrogen to a hydrogen.

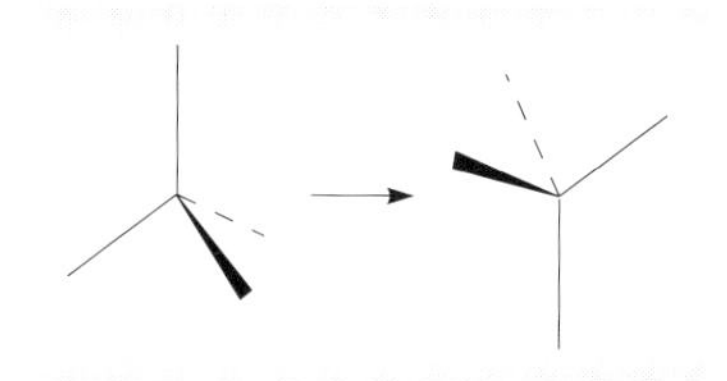

Figure 22 Inversion of NF_3 through the N atom.

Let us consider some other molecules. Does NF_3 have a centre of symmetry for example? No. There is only one nitrogen and so if there is a centre of symmetry it must be at the nitrogen. Inverting the fluorines through the nitrogen, however, turns the molecule upside down and left to right so that it does not look the same (Figure 22) and inversion is not a symmetry operation for NF_3.

How about BF_3: this had more axes and planes of symmetry than NF_3, so perhaps it has a centre of symmetry? No, the centre of symmetry would have to be the boron atom and, because there are an odd number of fluorines, at least one fluorine cannot have a partner lying the other side of the boron. In fact none of the fluorines has (Figure 23).

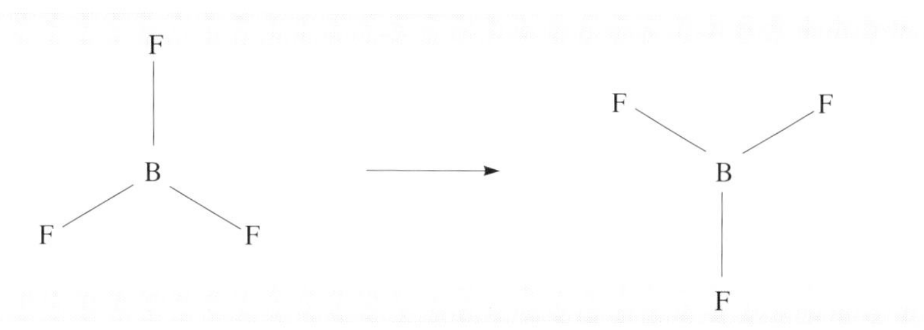

Figure 23 Inversion of BF_3 through the B atom.

So let's try something with an even number of atoms around a central atom, XeF_4. There is only one Xe atom and so, if there is a centre of symmetry, it must be at the Xe atom. A line from any one of the fluorines through the Xe atom and out the other side brings us to another fluorine, and so XeF_4 does have a centre of symmetry.

How about the other fluorides with four fluorines? In the tetrahedral molecules, such as CF_4, inversion through the carbon atom would turn the molecule inside out as in Figure 24. The molecule is not the same and so this is not a symmetry operation. Tetrahedral molecules do *not* have a centre of symmetry.

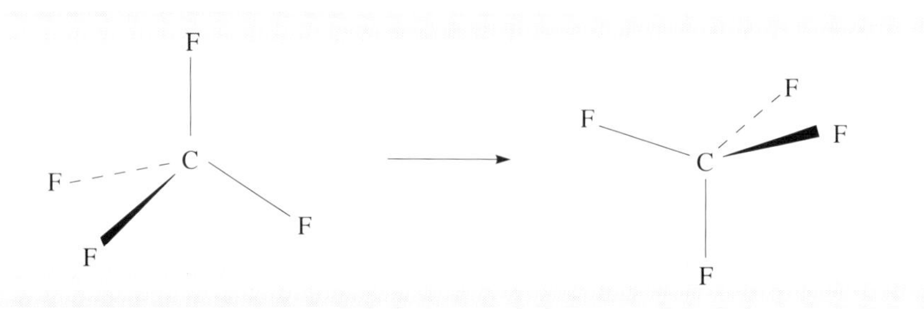

Figure 24 Inversion of CF_4 through the C atom.

In SF_4 the two axial fluorines inverted through the sulphur will change places but the equatorial fluorines would move to the other side of the molecule, thus changing its appearance. SF_4 does not have a centre of symmetry.

■ Which of the following molecules have a centre of symmetry: OF_2, F_2, HF, CO_2, and BrF_5?

□ F_2 and CO_2 only.

SAQ 14 Have the following molecules a centre of symmetry: (a) PF_5; (b) SF_6; (c) CO_3^{2-}; (d) ICl_2^- (Cl—I—Cl)$^-$; (e) ethene?

SAQ 15 The $CaCl_2$ molecule formed by heating solid calcium chloride has a centre of symmetry. Is it straight or V-shaped?

SAQ 16 A complex anion $[PtCl_4]^{2-}$ has a centre of symmetry. What shape is it?

5.4 SYMBOLS FOR SYMMETRY ELEMENTS

To save writing out axis of symmetry, plane of symmetry or centre of symmetry in full each time we mention them, we give each symmetry element a symbol. Thus, a plane of symmetry is given the symbol σ (sigma) and a centre of symmetry the symbol i. An *n*-fold axis of symmetry is given the symbol C_n; that is, a twofold axis is signified by C_2, a threefold axis by C_3, and so on. The corresponding symmetry operations (that is the acts of rotation, reflection, etc.) are given the same symbols, but we shall distinguish them by putting a circumflex accent over the symbol for an operation, for example $\hat{\sigma}$. Table 1 summarizes the symmetry elements and operations that you have met so far. If a molecule possesses more than one plane of symmetry or *n*-fold axis of symmetry, we can indicate this by putting the appropriate number before the symbol for the symmetry element. Thus, if the molecule has four twofold axes (for example XeF_4), we write $4C_2$. Vertical and horizontal planes of symmetry are distinguished by subscripts, σ_v and σ_h.

Let's finish this Section by listing all the symmetry elements of OF_2.

■ How many axes of symmetry has OF_2?

□ One twofold axis.

■ How many planes of symmetry has OF_2?

□ We saw in Section 5.2 that it had two.

■ Are these planes vertical or horizontal?

□ Vertical.

■ Does the molecule have a centre of symmetry?

□ No.

The complete list using the symbols just introduced is then $C_2 + 2\sigma_v$.

Table 1 Symbols for symmetry elements and symmetry operations

Symmetry element	Symbol	Symmetry operation	Symbol
plane of symmetry	σ, σ_v, σ_h	reflection through the plane	$\hat{\sigma}$, $\hat{\sigma}_v$, $\hat{\sigma}_h$
n-fold axis of symmetry	C_n	rotation through $1/n$ of a complete revolution about the axis of symmetry	$\hat{C}_n$
centre of symmetry	i	inversion through the centre of symmetry	$\hat{i}$

5.5 SUMMARY OF SECTIONS 5.1–5.4

1 The symmetry of a molecule or other object can be described by saying how many and which symmetry elements it contains.

2 Each symmetry element has associated with it a symmetry operation.

3 If an object contains a particular symmetry element, then the action of the associated symmetry operation will produce an identical-looking object occupying the same position in space as the original.

4 Important symmetry elements are a plane of symmetry, σ, an axis of symmetry, C_n, and a centre of symmetry, i.

5 The associated symmetry operations are reflection through the plane of symmetry, $\hat{\sigma}$, rotation by $1/n$ of a complete revolution about the axis of symmetry, $\hat{C}_n$, and inversion through the centre of symmetry, $\hat{i}$.

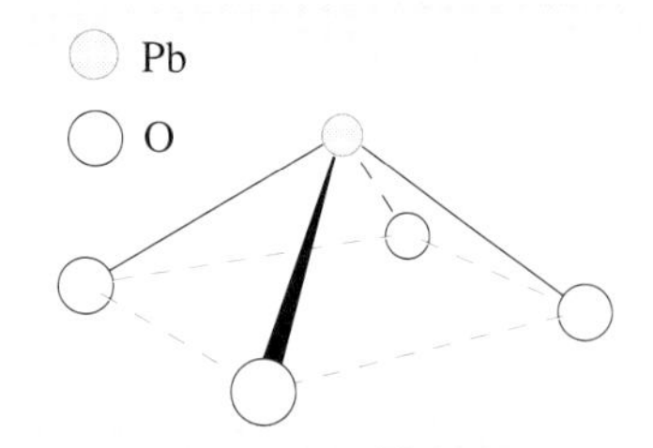

Figure 25 Pb^{2+} in a crystal of PbO.

SAQ 17 List all the symmetry elements of the molecules (a) BF_3, (b) SF_4, and (c) IF_7, and give them their correct symbols.

SAQ 18 Figure 25 shows the environment of a Pb^{2+} ion in a crystal of PbO. List all the symmetry elements you can find.

SAQ 19 Find the symmetry elements in the following: (a) a spoon; (b) a 50 pence coin (ignore the designs on the head and tail); (c) a saucer; (d) a snowflake; (e) a hand.

5.6 SYMMETRY POINT GROUPS

We have looked at several molecules to see what symmetry elements they contain. Now we are going to meet a way of classifying molecules (and other objects) according to their symmetry. What we do is to group together all molecules and other objects that contain the same symmetry elements. The symmetry operations of any molecule form a mathematical group called a **symmetry point group**. All molecules with the same symmetry elements will be unchanged by the same symmetry operations and hence are said to belong to the same symmetry point group, denoted by a particular symbol.

OF_2 and all objects that contain one twofold axis of symmetry and two vertical planes of symmetry belong to the same group, denoted by the symbol $\mathbf{C_{2v}}$.

■ Which other fluorides among your models belong to the symmetry point group $\mathbf{C_{2v}}$?

□ SF_4 and ClF_3 both belong to $\mathbf{C_{2v}}$. Their symmetry elements are shown in Figure 26. Note that molecules in the same symmetry point group do not necessarily look similar, even though they contain the same symmetry elements.

You should check for yourself that these molecules contain the same symmetry elements as OF_2, *and no others*.

Let us now look at another group, the group of molecules with the same symmetry elements as NF_3 (Figure 19).

■ What symmetry elements does NF_3 contain?

□ One threefold axis and three vertical planes of symmetry.

The group of molecules with the same symmetry elements as OF_2 was labelled $\mathbf{C_{2v}}$ because the axis of highest symmetry was twofold, and the two planes of symmetry were vertical.

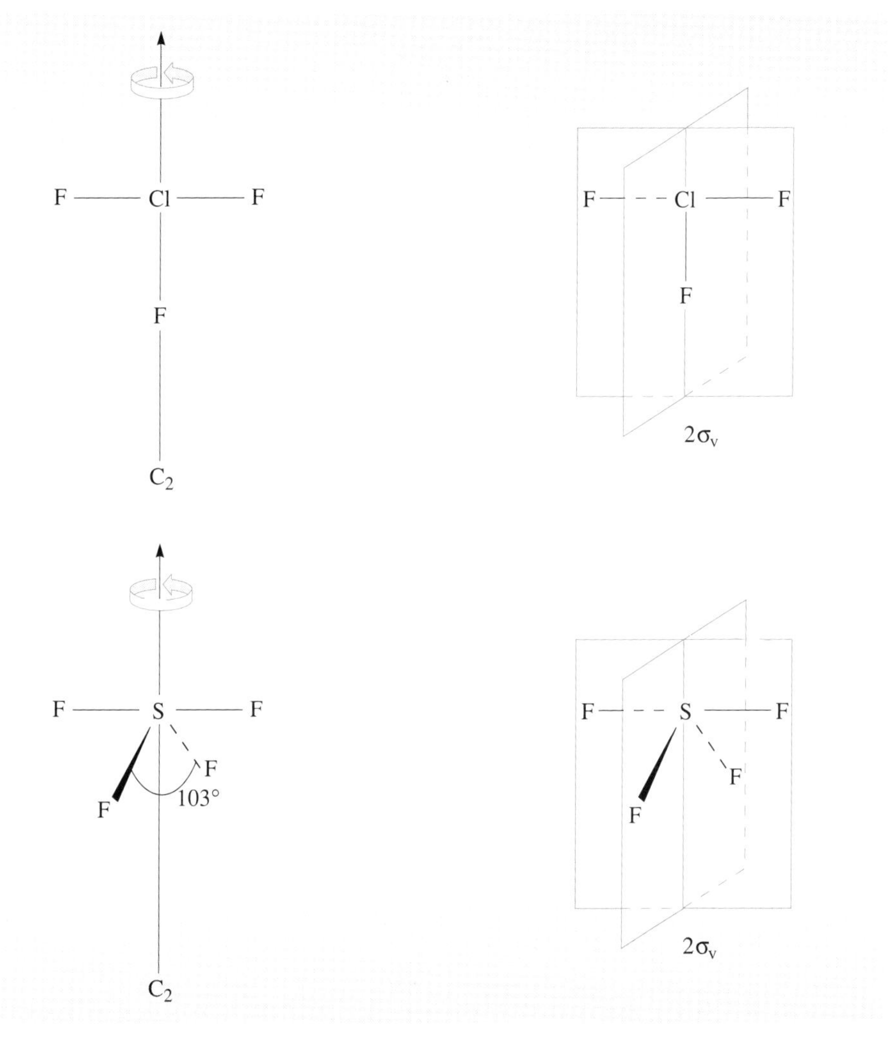

Figure 26 The symmetry elements of ClF_3 and SF_4.

■ Can you suggest a symbol for the group of molecules with the same symmetry elements as NF_3?

□ The symmetry point group is labelled $\mathbf{C_{3v}}$.

The members of the group $\mathbf{C_{2v}}$ have symmetry elements $C_2 + 2\sigma_v$. Those in $\mathbf{C_{3v}}$ contain the elements $C_3 + 3\sigma_v$. What symmetry elements would you expect molecules in a group labelled $\mathbf{C_{4v}}$ to contain? By comparison with $\mathbf{C_{2v}}$ and $\mathbf{C_{3v}}$, you probably said $C_4 + 4\sigma_v$, and this is correct. Thus, we have a whole set of groups $\mathbf{C}_{n\mathrm{v}}$ whose members contain the symmetry elements $C_n + n\sigma_v$. We can even have the group for which n is infinite, namely $\mathbf{C_{\infty v}}$. This group is important because it contains all diatomic molecules in which the two atoms are different (**heteronuclear diatomic molecules**), such as HF. It also contains all linear molecules that do not have a centre of symmetry, for example HCN.

Are there any other sets of groups?

■ What are the symmetry elements of BF_3?

□ $C_3 + 3C_2 + 3\sigma_v + \sigma_h$. Look at your model.

The axis of highest symmetry is the C_3 axis, but BF_3 does not belong to the group $\mathbf{C_{3v}}$ because it contains more symmetry elements than the molecules in that group.

■ What are the extra symmetry elements in BF_3?

□ $3C_2 + \sigma_h$.

The group to which BF_3 belongs is labelled $\mathbf{D}_{3h}$. D_3 is used rather than C_3 for a threefold axis when a molecule has three twofold axes at right-angles to the threefold axis. The subscript h stands for horizontal, and is added if there is a horizontal plane of symmetry.

The symmetry elements of the group $\mathbf{D}_{3h}$ are therefore $C_3 + 3\sigma_v + 3C_2 + \sigma_h$. This is a representative of another set of groups $\mathbf{D}_{nh}$, which contain the elements $C_n + n\sigma_v + nC_2 + \sigma_h$. If n is an even number or infinity, the molecule also possesses a centre of symmetry, i.

■ What is the highest axis of symmetry in XeF_4?

□ A fourfold axis, C_4.

■ Are there four twofold axes perpendicular to this C_4 axis?

□ Yes. Look at your model. The two lines passing along two Xe—F bonds are C_2 axes, as are the two lines bisecting the FXeF angles.

■ Can you suggest which symmetry point group XeF_4 might belong to?

□ Since it has four C_2 axes at right-angles to its C_4 axis it does *not* belong to $\mathbf{C}_{4v}$. It might, however, belong to $\mathbf{D}_{4h}$.

■ What symmetry elements would a molecule in $\mathbf{D}_{4h}$ have?

□ $C_4 + 4C_2$ (perpendicular to the C_4) $+ 4\sigma_v + \sigma_h + i$.

■ Does XeF_4 contain all these elements?

□ Yes. Use your model if necessary to convince yourself of this.

As for $\mathbf{C}_{nv}$, n in $\mathbf{D}_{nh}$ can be infinite. The group $\mathbf{D}_{\infty h}$ contains all diatomic molecules with two identical atoms, for example F_2, N_2, and all linear molecules with a centre of symmetry, such as BeF_2 and CO_2.

The two sets of groups $\mathbf{C}_{nv}$ and $\mathbf{D}_{nh}$ are among the most important for chemists, because molecules belonging to them are quite common.

There are two other important groups that we have not yet mentioned but which are also common. These are the groups of objects with the same symmetry elements as a tetrahedron and as an octahedron. Molecules and other objects with the same symmetry elements as a tetrahedron, for example CF_4 and SiF_4, are labelled $\mathbf{T}_d$ (T for tetrahedron and d for dihedral*). Octahedral molecules, for example SF_6, belong to the symmetry point group $\mathbf{O}_h$ (O for octahedron and h for horizontal). We are not going to look at these groups in detail. They contain a large number of symmetry elements. It is usually very easy to see whether a molecule is octahedral or tetrahedral, but note that only regular tetrahedral and octahedral molecules belong to $\mathbf{T}_d$ and $\mathbf{O}_h$, respectively. Thus CF_4 belongs to $\mathbf{T}_d$ but CF_3H does not. When discussing VSEPR we often referred to a molecule with four repulsion axes as adopting a tetrahedral arrangement, even if the four repulsion axes were not identical. To belong to $\mathbf{T}_d$, however, the molecule or ion must have four identical bonds arranged tetrahedrally.

* Dihedral symmetry planes will not be considered further in this Course. Molecules in the group $\mathbf{T}_d$ do not have a unique principal axis but have several axes of the same order. The σ_d planes each contain one of these axes but neither contain nor are at right-angles to the others.

Some of the molecules we have met belong to other groups. For example, HOF has only one symmetry element, a plane of symmetry. It belongs to the group $\mathbf{C}_s$. Molecules with only an axis of symmetry belong to a group $\mathbf{C}_n$, where the axis is n-fold. Molecules whose only symmetry element is a centre of symmetry belong to the group $\mathbf{C}_i$ and those with no symmetry elements belong to $\mathbf{C}_1$.

There is one more family of groups that we want to introduce here, and that is the set $\mathbf{C}_{nh}$. These possess a $\mathbf{C}_n$ axis, a horizontal plane of symmetry and, if n is even, a centre of symmetry. There are a fair number of molecules belonging to $\mathbf{C}_{2h}$, for example *trans*-1,2-difluoroethene, but molecules belonging to $\mathbf{C}_{nh}$ with n greater than 2 are relatively rare. This molecule belongs to $\mathbf{C}_{3h}$:

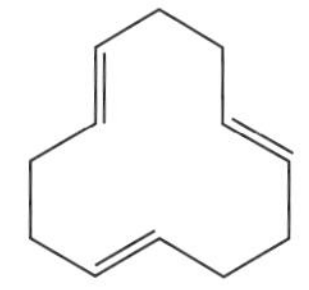

By looking at the key symmetry elements of a molecule, we can assign it to a symmetry point group. We have written the rules as a sort of flow chart. This is printed on the fold-out page at the back of the Block. To find the point group of a particular molecule, start at the top of the page and answer the questions as you work down. Let's see how this chart works with a molecule whose symmetry point group we know, BF_3.

■ First we have to decide if BF_3 is tetrahedral.

□ The answer to this is no, so we follow the no line to the next question.

■ Is BF_3 octahedral?

□ No.

■ Has BF_3 a C_n axis?

□ Yes.

■ What is the largest value of n?

□ Three. This is the value of n you must now use in all subsequent questions in the chart for this molecule.

■ Are there three C_2 axes perpendicular to the C_3 axis?

□ Yes.

■ Is there a plane of symmetry perpendicular to the C_3 axis?

□ Yes.

So our chart tells us that BF_3 belongs to the point group $\mathbf{D}_{3h}$, which is correct. Let's use the flow chart to find the symmetry point groups to which a few more of the molecules you have met belong.

First let's try BrF_5. This is neither tetrahedral nor octahedral. Although it has six repulsion axes, one of these is a lone pair and so BrF_5 is not a regular octahedral molecule and does not belong to $\mathbf{O}_h$.

■ Has BrF_5 any axes of symmetry?

□ Yes, it has one, C_4.

As there is only one axis of symmetry, this must be the axis of highest symmetry, so $n = 4$.

■ Are there four C_2 axes perpendicular to the C_4 axis?

□ As there is only one axis and this is C_4, the answer is obviously no.

■ Does BrF_5 have four vertical planes of symmetry?

□ BrF_5 has four planes of symmetry. They all contain the C_4 axis and are thus vertical.

BrF_5 therefore belongs to the group $\mathbf{C}_{4v}$.

Finally let's try PF_5. PF_5 is neither tetrahedral nor octahedral.

■ Does it have any axes of symmetry?

□ Yes. It has one threefold axis and three twofold axes. Look at your model.

■ The largest value of n is three. Does PF_5 have three C_2 axes perpendicular to the C_3 axis?

□ Yes. The three C_2 axes are along the equatorial P—F bonds and are at right-angles to the C_3 axis, going through the axial P—F bonds.

■ Is there a plane of symmetry perpendicular to the C_3 axis?

□ Yes: the plane of the three equatorial P—F bonds.

So PF_5 belongs to $\mathbf{D}_{3h}$, the same group as BF_3.

5.7 SUMMARY OF SECTION 5.6

1 All objects with the same symmetry elements belong to the same symmetry point group.

2 By following the rules in the flow chart you can assign a given molecule to its symmetry point group.

3 The most important point groups for chemists are $\mathbf{C}_{nv}$, $\mathbf{D}_{nh}$, $\mathbf{T}_d$ and $\mathbf{O}_h$.

SAQ 20 Use your flow chart to determine the symmetry point groups of the following molecules: (a) SO_2 (V-shaped); (b) KrF_2; (c) carbon monoxide, CO; (d) $POCl_3$ (the O atom and the three Cl atoms are arranged approximately tetrahedrally around the central P atom); (e) NSF (bent).

SAQ 21 Use your flow chart to determine the symmetry point group of the following objects: (a) a saucer; (b) a spoon; (c) a snowflake; (d) a starfish.

SAQ 22 Below are two lists. The first contains a number of molecules and the second some symmetry point groups. Match each molecule in the first list with a point group in the second list.

Molecule	Symmetry point group
(a) H_2	(i) $\mathbf{C}_{2v}$
(b) ethene	(ii) $\mathbf{C}_{3v}$
(c) trifluoromethane, CHF_3 (like CF_4 with one F replaced by H)	(iii) $\mathbf{C}_{\infty v}$
(d) O_3 (bent)	(iv) $\mathbf{D}_{2h}$
(e) HCl	(v) $\mathbf{D}_{\infty h}$

SAQ 23 Which of the following symmetry point groups contain molecules that have a centre of symmetry: $\mathbf{D}_{2h}$, $\mathbf{C}_{5v}$, $\mathbf{C}_{i}$, $\mathbf{D}_{3h}$, $\mathbf{C}_{6v}$?

SAQ 24 Show that CF_4, CF_3Cl and CF_2Cl_2, although all based on a tetrahedral shape according to VSEPR, belong to different symmetry point groups by considering the number of threefold axes in each molecule.

STUDY COMMENT Video sequence 6, entitled *An exercise in symmetry determination*, gives you more opportunity to practise your skills in this area. You should watch it soon, and also use it for revision purposes.

6 BEYOND THE ELECTRON-PAIR BOND

The electron-pair model of bonding and VSEPR theory have the virtue of simplicity. They are easy to apply and to picture and, for many purposes, they provide an adequate description of molecules or at least a good starting point for more sophisticated theories.

They do, however, have shortcomings. We cannot, for example, use them to predict bond lengths. The very large class of transition metal compounds, which you will meet if you go on to third level chemistry, cannot be described using VSEPR. For example, the ions VCl_4^-, $MnCl_4^{2-}$, $FeCl_4^{2-}$ and $CoCl_4^{2-}$ are all tetrahedral although they have different numbers of valence electrons. Even for such a simple molecule as dioxygen, O_2, we need to go to a different theory to explain why it has two unpaired electrons.

The theory that most chemists today use is molecular orbital theory, and in the following sections we are going to introduce some of the basic ideas of this theory.

STUDY COMMENT Molecular orbital theory is also described in video sequence 7, *A new look at bonding*. Ideally, you should watch this during your study of Section 6.

As in the electron-pair bond, the atoms in a molecule are held together by sharing electrons, but molecular orbital theory enables us to be more specific about how they are shared.

The theory is based on the concept of an electron behaving as a wave, and one of the consequences of this is that we cannot picture the electron as moving round the nuclei in a definite path. We can, however, say how likely the electron is to be at any particular point. When we observe a molecule experimentally, for example by X-ray diffraction, there is a higher density of electrons at some points than at others. A way of picturing this is to imagine the electrons smeared out over the atom or molecule, more thickly spread in some places than others. In Figure 27 for example, we have pictured an electron in a hydrogen atom; the areas where the dots are closer together are those where the electron is more likely to be found (or is more thickly spread, if you like).

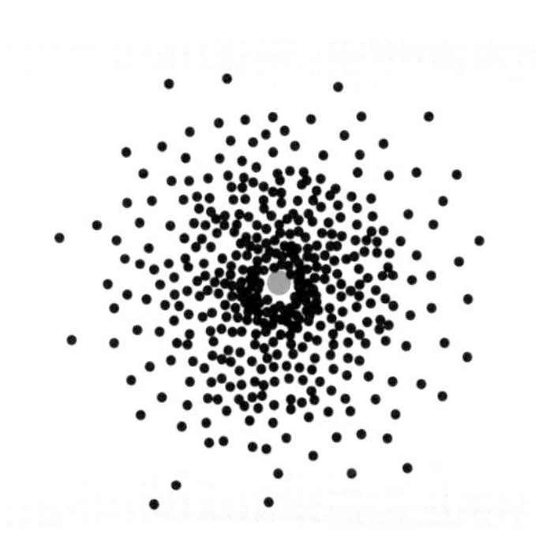

Figure 27 Imaginary photographs of an electron in a hydrogen atom.

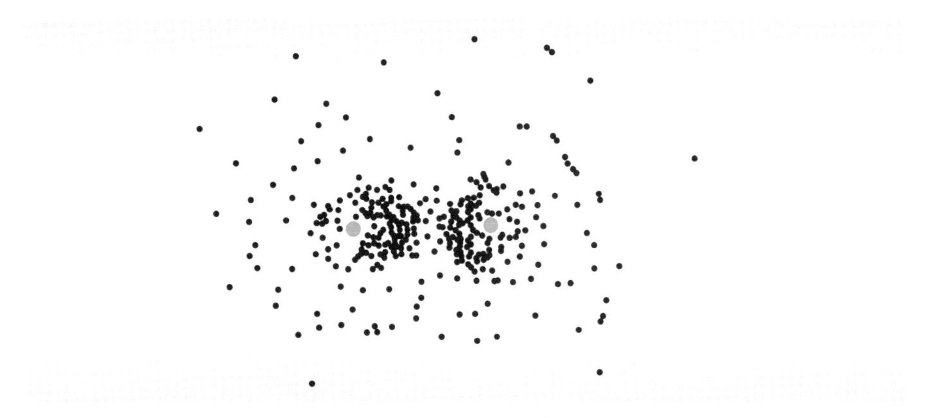

Figure 28 Imaginary photographs of an electron in H_2^+.

If we now add another hydrogen nucleus to form H_2^+, the electron distribution changes. The electron now is likely to be near both nuclei and is most likely to be between the two (Figure 28).

Molecular orbital theory enables us to calculate a **wavefunction** for electrons in molecules and, from this, the energy and other properties of molecules. The probability of an electron being at any point is, for example, given by the square of the wavefunction. However, although it is possible to write down a mathematical equation whose solution is the electron wavefunction for any molecule, in practice it is not possible to solve the equation exactly except for small and chemically uninteresting molecules such as H_2^+. Chemists have therefore developed approximate solutions, some of which give answers that agree remarkably well with experiments.

The major approximation made is that the description of an electron in a molecule can be related to the descriptions of electrons in the constituent atoms. First then we shall look at electrons in atoms.

6.1 ATOMIC ORBITALS

An electron in an atom is subject to two influences. Because it is charged, it is attracted to the (oppositely charged) nucleus and repelled by any other electrons present. In addition it is moving around, and this prevents it falling into the nucleus (much as the Earth, although attracted to the Sun, does not fall into it). What we find is that there are certain electron distributions for which these two influences balance. If we work out the energies of these distributions for the hydrogen atom, we find they correspond to the experimentally found energy levels 1s, 2s, 2p, 3s, 3p, 3d, etc. The distribution in Figure 27 is that of an electron in the 1s level.

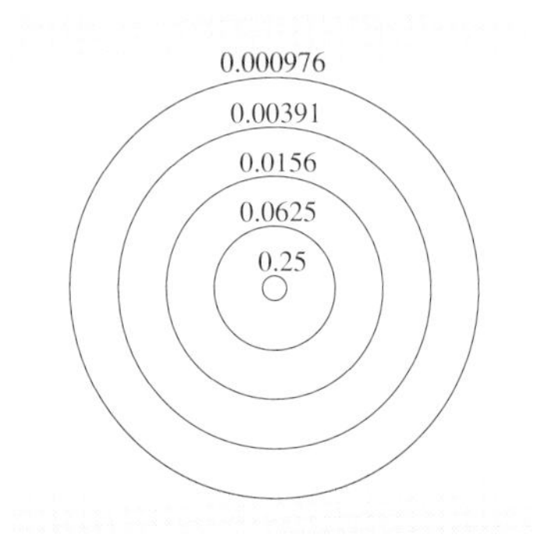

Figure 29 Contour diagram for a 1s electron in a hydrogen atom. The numbers denote electron density.

A common way of representing such distributions is by means of a contour diagram. In this, we draw a line joining all points where the electron is equally likely to be. Such a diagram is shown for the 1s orbital in Figure 29.

A quicker way of representing the electron is to pick just one contour. The one chosen is usually one such that the electron has a 95% chance of being inside it. Figure 30 illustrates this contour.

Finally, now that there are computer programs that will draw fancy pictures, three-dimensional representations are becoming common. One such is shown for the 1s electron in Figure 31. Here the height above the plane represents the likelihood of the electron being at a point in the plane. In general, we shall use the single contour as in Figure 30 because this is more easily remembered and drawn freehand.

The 95% probability contours for electrons of 2s, 2p, 3s, 3p and 3d energy levels in hydrogen are shown in Figure 32. Note that the distribution of the s electrons is spherical, that of the p electrons forms two separate lobes, and that of the d electrons forms four lobes.

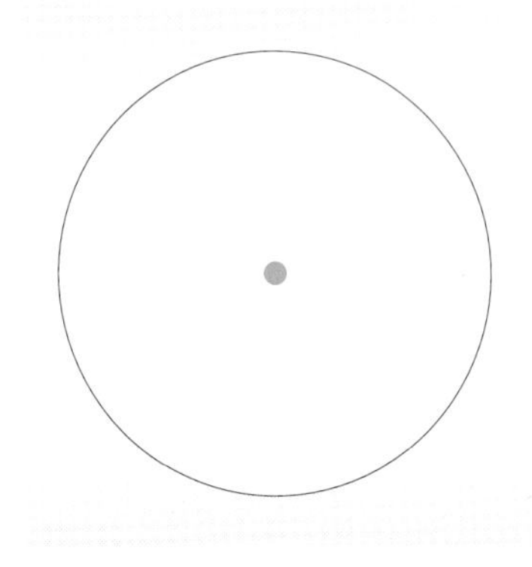

Figure 30 The 95% probability contour for a 1s electron in a hydrogen atom.

Figure 31 Computer plot of a 1s electron in a hydrogen atom. The probability of being at any point in a plane is represented by a height above the plane. The peak is above the nucleus.

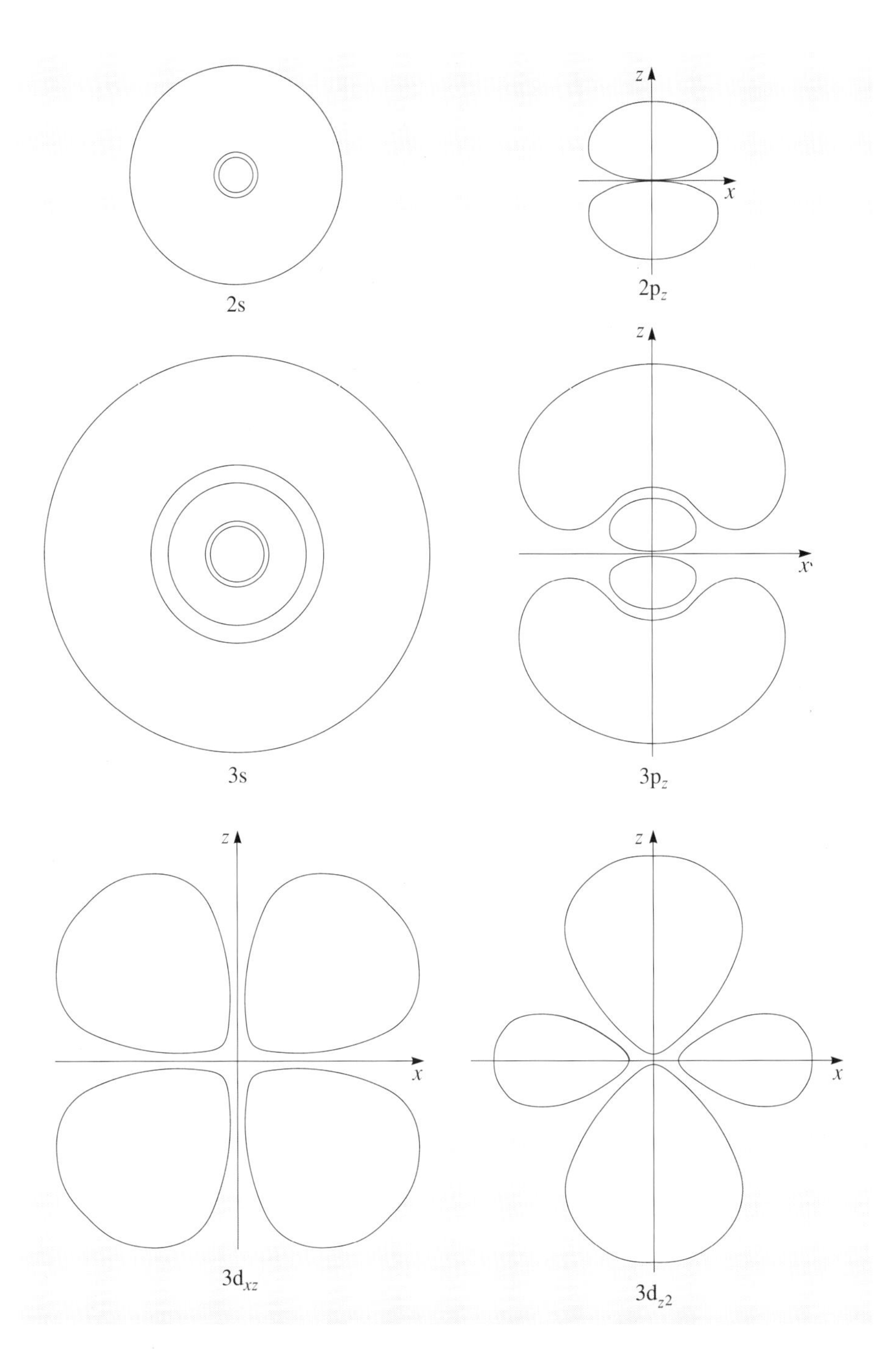

Figure 32 The 95% probability contours for 2s, 2p, 3s, 3p and 3d electron distributions in the hydrogen atom.

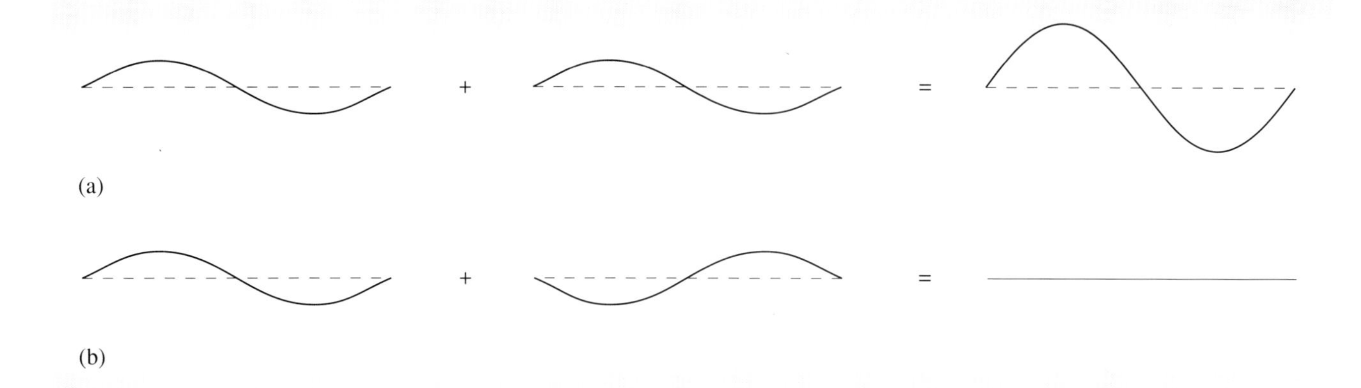

Figure 33 Simple waves.

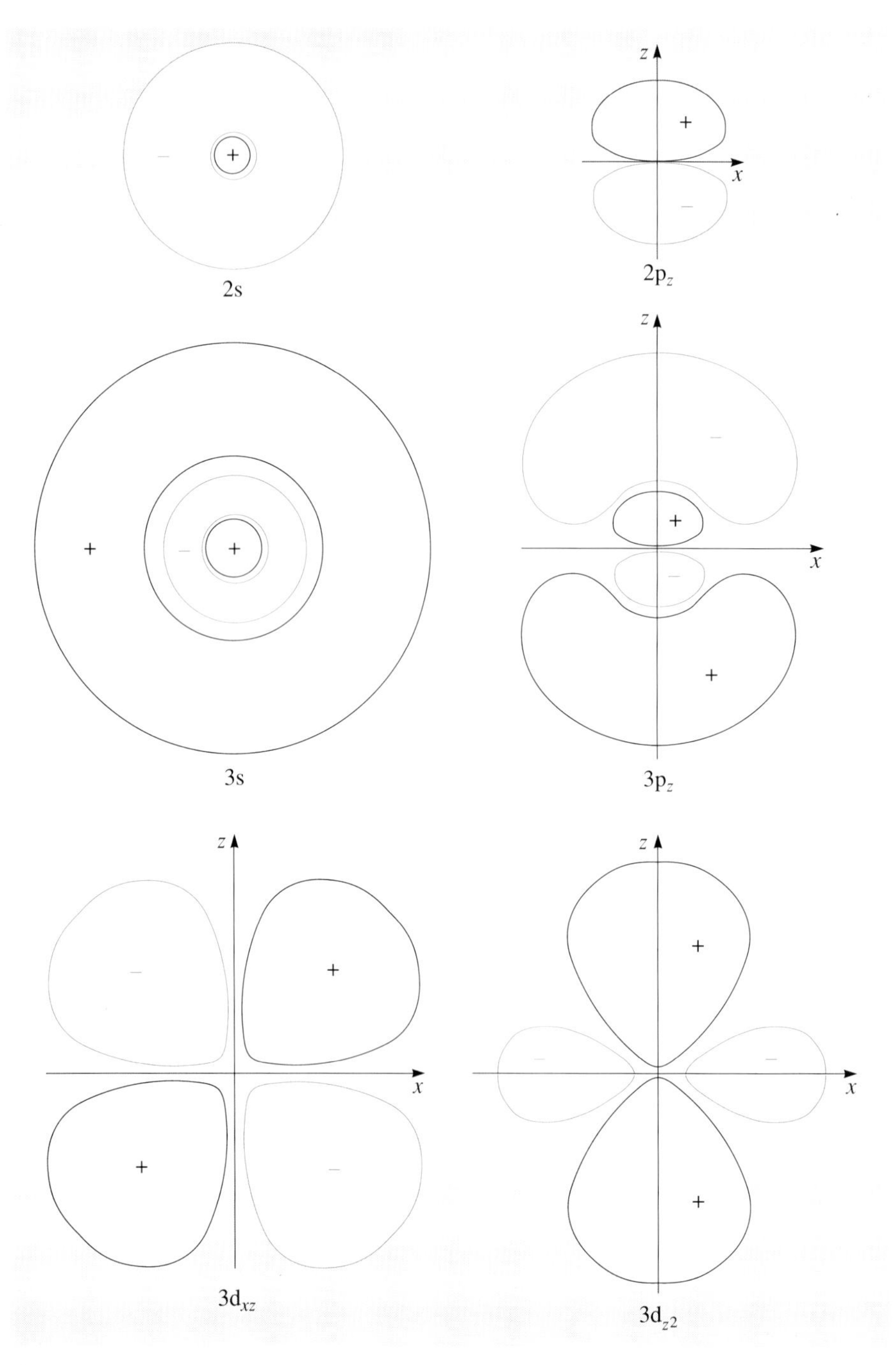

Figure 34 2s, 2p, 3s, 3p and 3d orbitals of the hydrogen atom, with the phases denoted by + and – signs.

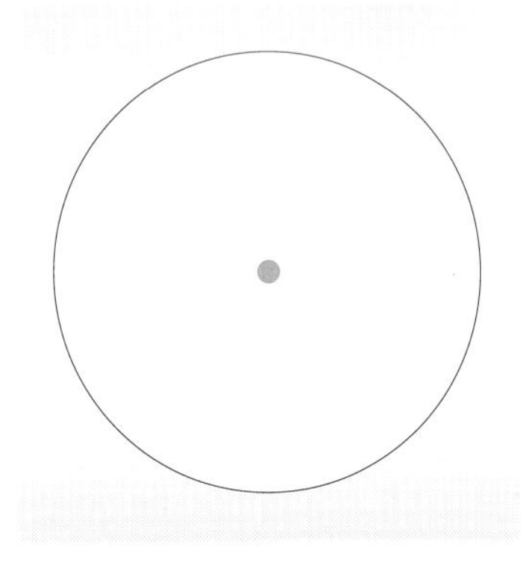

The figures we have drawn so far show only how much electron there is at any point, but when we come to combine atoms to form molecules, another property of the electron wave is also important. This is the phase or sign. Figure 33 shows some simple waves. In Figure 33a the two waves go up and down at the same time: they are said to be **in phase**. If we add these two waves together then we get a bigger wave, which is still moving up and down in the same pattern. In Figure 33b one wave goes up as the other goes down: the waves are **out of phase**. If these two waves are added together, they cancel each other out.

The wavefunctions describing electrons in atoms or molecules are called **orbitals**. They are more complex than the simple waves in Figure 33, but when they combine in phase or out of phase the results are similar.

The electron distributions in Figure 32 are given by the square of the wavefunction. To represent the wavefunction itself we can use the same contours, but we also need to indicate the phase of the wavefunction. On atomic and molecular orbital representations, we shall use colour to show differences in phase. 1s orbitals are all one phase and so are shown in one colour. 2p orbitals have two parts, which are out of phase with each other, and we illustrate this by showing one part in black and one in colour. The four lobes of the 3d orbital alternate in phase.

Figure 34 shows the same orbitals as in Figure 32 but coloured to indicate parts of the orbital that differ in phase. The different phases can also be denoted by + and – signs, and these are also shown in Figure 34. The + and – refer to the value of the wavefunction; in the regions labelled + the mathematical function describing the electron wave is a positive (fractional) number, whereas in the regions labelled – it is a negative number. (Note that + and – do *not* represent electrical charge.)

6.1.1 ORBITALS OF ATOMS OTHER THAN HYDROGEN

The orbitals in Figure 34 can be calculated exactly because hydrogen has only one electron. With more than one electron we have to balance not only the motion and nuclear attraction of each electron but also the repulsion between electrons. Even with today's large, fast computers, the exact wavefunctions of atoms cannot be calculated and so approximations have to be made. What is generally done is assume that each electron is in an orbital like the hydrogen orbitals. So for helium we put two electrons of opposite spin in a 1s orbital, for lithium two electrons go into 1s and one into 2s, for carbon two electrons go into 1s, two into 2s and two into 2p orbitals.

These orbitals are not the same as the corresponding orbitals in hydrogen. Because all atoms are spherical, the orbitals can still be labelled s, p, d, etc., but the average distances of the electrons from the nucleus are different from those in hydrogen. Figure 35 shows the 1s orbital in lithium compared with that in hydrogen.

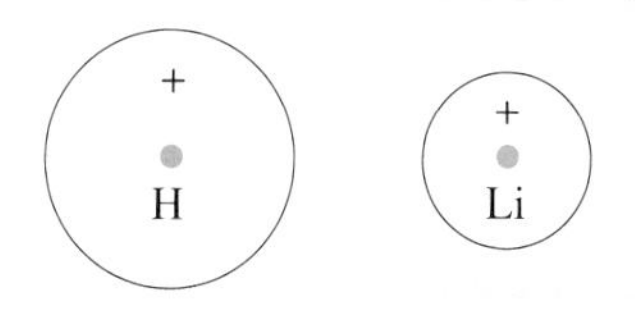

Figure 35 1s orbitals in H and Li.

Suppose we wanted to calculate an orbital in lithium. We would start off by putting one electron in a 2s orbital and two in a 1s orbital, using a guess as to what the orbitals were like. Then one of the orbitals is recalculated assuming that the electron in it is influenced by the nucleus and a cloud of negative charge produced by the other two electrons. We then place an electron in the recalculated orbital and recalculate one of the others. This process continues until recalculation no longer changes the orbital. The final orbital is for an electron that feels the nuclear charge through a screen of other electrons.

Having obtained the atomic orbitals, we then assign electrons to them starting from the lowest energy orbital. Two electrons, of opposite spin, can go to into each orbital and, where there are two or more orbitals with the same energy, electrons go into separate orbitals with parallel spins first (Hund's rule).

So going across the second row of the Periodic Table, in lithium we put two electrons into 1s and one into 2s ($1s^22s$), in beryllium we put two electrons into each of 1s and 2s, at boron we add an electron to a 2p orbital ($1s^22s^22p$). At carbon we have to put two electrons into 2p orbitals. There are three 2p orbitals all of the same energy, but electrons with parallel spins tend to keep apart and so putting the 2p electrons in with parallel spins reduces electron repulsion. Rather than put two electrons of opposite spin into one 2p orbital, therefore, we place the electrons one in each of two 2p orbitals with parallel spins. For nitrogen we have two paired electrons in 1s, two paired electrons in 2s and three electrons with parallel spins in the three 2p orbitals. At oxygen we have to start pairing 2p electrons so we have a pair in 1s, a pair in 2s, a pair in one of the 2p orbitals and two single electrons with parallel spins in each of the other two 2p orbitals. Fluorine has two paired 1s electrons, two paired 2s, two 2p orbitals each with a pair of electrons in and one 2p orbital with one electron in. Finally at neon the 1s, 2s and all three 2p orbitals contain pairs of electrons.

6.2 ORBITALS IN DIATOMIC MOLECULES

An electron in a diatomic molecule is attracted to the two nuclei of the molecule. Ideally we should find new orbitals that take account of this. Exact orbitals of this type are very difficult to obtain, and so we use the idea that molecules are combinations of atoms. This means that we combine atomic orbitals to make **molecular orbitals**.

Take the simplest molecule, H_2^+. This has two hydrogen nuclei and one electron. We need to find an orbital that will describe this electron. Since an electron in a hydrogen atom is in a 1s orbital, we make an attempt to approximate the molecular orbital by combining 1s orbitals. The usual way to combine atomic orbitals to make molecular orbitals is to add them as we added the simple waves in Figure 33. So we might try adding together a 1s orbital on one hydrogen nucleus and a 1s orbital on the other hydrogen nucleus. The two 1s orbitals are shown in Figure 36. If the two orbitals are in phase as shown in the figure, then where they overlap, the wavefunction increases. This represents an increase of **electron density**. The overlap lies between the two nuclei and so, in a combination like this, there would be a high density of electron between the nuclei. Now this looks useful. If the electron is concentrated between the nuclei, it can act to hold the nuclei together. An electron in such an orbital would be effectively binding the two hydrogen nuclei together to make a molecule.

Combining two 1s orbitals in this manner, then, seems a good first approximation to describe bonding using this theory.

We shall just define this molecular orbital a little more closely. The molecular orbital describes only one electron. The density of electron at any point in space is given by the square of the electron wavefunction. If I sum this density over all space, I should get an answer of one because the total density for any orbital corresponds to just one electron. If I simply add two 1s orbitals, then the total density will be given by $(1s_A + 1s_B)^2$, where $1s_A$ represents a 1s orbital on one nucleus and $1s_B$ a 1s orbital on the other nucleus:

$$(1s_A + 1s_B)^2 = 1s_A^{\ 2} + 1s_B^{\ 2} + 2(1s_A \times 1s_B)$$

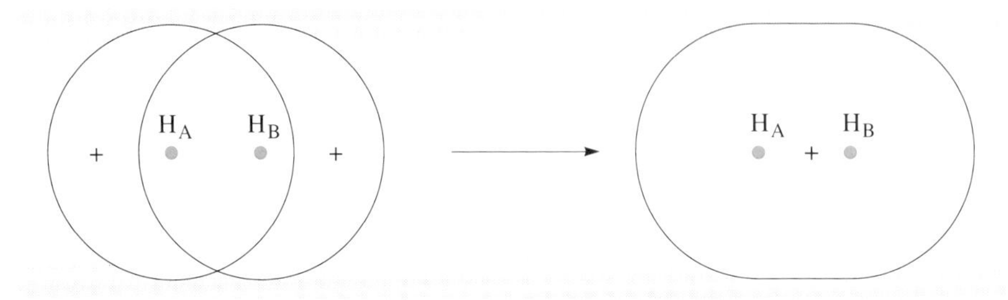

Figure 36 Overlap of two in-phase 1s orbitals in H_2^+.

Now the total density for a 1s orbital is also one and so the sum of $1s_A{}^2$ over all space is one. So is the sum of $1s_B{}^2$. The total density for the molecular orbital as it stands is thus 2 + twice $1s_A1s_B$ summed over all space, which gives a value greater than one. We therefore add together only a fraction of each 1s orbital and our molecular orbital is written $N(1s_A + 1s_B)$, where the sum of $N^2(1s_A + 1s_B)^2$ over all space is one.

Will the electron in $H_2{}^+$ occupy the orbital $N(1s_A + 1s_B)$ rather than a 1s orbital on one H atom? Yes it will, because an electron in $N(1s_A + 1s_B)$ is more likely to be attracted to both nuclei than one in a 1s orbital. This means that the electron is more strongly bound and its energy is lower. An electron in $N(1s_A + 1s_B)$ is therefore of lower energy than one in a 1s orbital. Orbitals such as $N(1s_A + 1s_B)$ are called **bonding orbitals.**

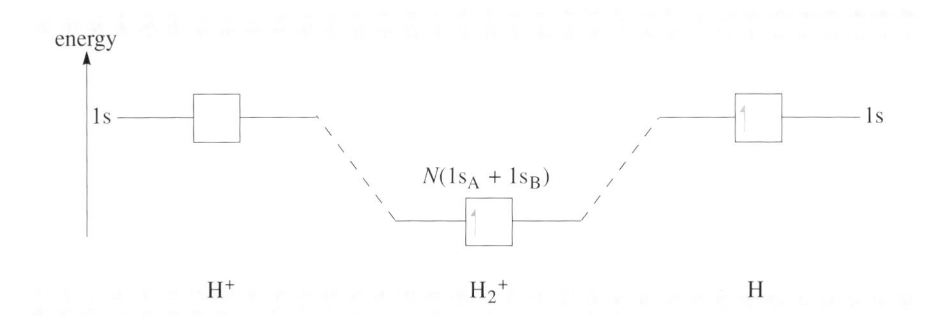

Figure 37 Energy-level diagram for $H_2{}^+$.

We can draw a diagram (Figure 37) to illustrate the energy advantage of our bonding orbital. In the centre we put boxes or lines to represent molecular orbitals. Either side, or with more complicated molecules on one side, we put boxes to represent the orbitals on the atoms that make up the molecule. Energy is represented as increasing up the page. Dashed lines join the molecular orbitals to the atomic orbitals that are combined to make them. So, in the centre of Figure 37 we have $H_2{}^+$ and the orbital $N(1s_A + 1s_B)$. Either side we have 1s orbitals. The 1s orbitals are shown higher up the page than $N(1s_A + 1s_B)$ because they are higher in energy, and lines join them to $N(1s_A + 1s_B)$ because the molecular orbital is a combination of 1s orbitals.

Figure 37 shows that we can consider $H_2{}^+$ as being formed from a hydrogen atom, H, and a hydrogen ion, H^+. To represent the electron in the atom, we place a half-headed arrow in the 1s box. The 1s box for the ion is left empty because H^+ has no electrons. $H_2{}^+$ has one electron and this goes into the box $N(1s_A + 1s_B)$.

$H_2{}^+$ is easy to deal with because it has only one electron, but it is not a very interesting molecule. Can we use the idea of combining atomic orbitals when we come to study molecules with more than one electron? Let us start with H_2, dihydrogen.

H_2 has two electrons and we can use a building-up process similar to that used to determine electron configurations for atoms. The orbital $N(1s_A + 1s_B)$ can take two electrons of opposite spin. If we draw a molecular orbital diagram for H_2 we put one electron in each 1s box and two paired electrons in the $N(1s_A + 1s_B)$ box, as shown in Figure 38. The two hydrogen atoms in H_2 are bonded together by a pair of electrons in a bonding orbital. We have an electron-pair bond.

We should point out here that the bonding orbital $N(1s_A + 1s_B)$ will not be exactly the same for $H_2{}^+$ and H_2. As with atoms, when we have more than one electron we have to allow for the other electrons when we calculate an orbital. So, for H_2, we would place

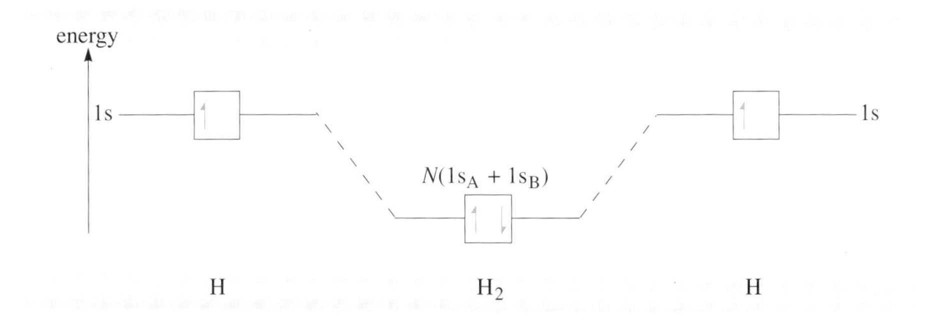

Figure 38 Energy-level diagram for H_2.

one electron in a trial $N(1s_A + 1s_B)$ and calculate a new $N(1s_A + 1s_B)$ for the other electron. We then place an electron in the new $N(1s_A + 1s_B)$ and recalculate, continuing until there is no change in $N(1s_A + 1s_B)$.

We can use Figure 38 to explain why hydrogen gas is found as H_2 molecules rather than as hydrogen atoms. Consider the process

$$H_2(g) \longrightarrow 2H(g)$$

As hydrogen is found as H_2 rather than as atoms, $\Delta G_m^\ominus$ for this process must be large and positive. $\Delta G_m^\ominus$, as you know, is the difference of two terms, $\Delta H_m^\ominus$ and $T\Delta S_m^\ominus$. $\Delta S_m^\ominus$ for this reaction will be large and positive because we are increasing the number of moles of gaseous species (Block 1, Section 2.6), and so for $\Delta G_m^\ominus$ to be large and positive, $\Delta H_m^\ominus$ must be positive and larger than $T\Delta S_m^\ominus$. Suppose we ionize H_2 and two hydrogen atoms until we are left with bare protons and free electrons. We can write a thermodynamic cycle for the equation above involving this ionization process. All substances in this cycle (Figure 39) are in the gas phase.

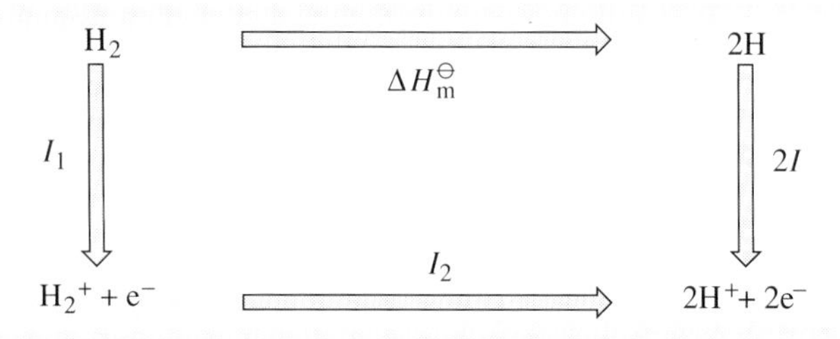

Figure 39 Thermodynamic cycle for the reaction $H_2 \longrightarrow 2H$.

The first ionization energy of H_2 is equal to minus the orbital energy. Now the orbital energy in H_2 is less than the atomic orbital energy, so I_1 for the dihydrogen molecule will be *greater* than I for the hydrogen atom. Once one electron has been removed from H_2 , the second will be harder to remove, so I_2 will be larger than I_1. The sum $I_1 + I_2$ will therefore be greater than $2I$, and because $\Delta H_m^\ominus$ is given by $(I_1 + I_2 - 2I)$, $\Delta H_m^\ominus$ is positive.

■ Would you predict H_2 to be stable with respect to two hydrogen atoms?

□ Yes, so long as the difference in ionization energies $(I_1 + I_2 - 2I)$ is sufficient to overcome the entropy term. The entropy term $T\Delta S_m^\ominus$ for this and similar reactions at room temperature is about 30 kJ mol^{-1}, and so $(I_1 + I_2 - 2I)$ must be greater than 30 kJ mol^{-1}. The value of I is 1 314 kJ mol^{-1} (*Data Book*), and that of I_1 is 1 574 kJ mol^{-1}. Because we are removing an electron from a positive ion, I_2 is clearly larger than I_1, and so $\Delta H_m^\ominus$ will be larger than $2(I_1 - I) = 520$ kJ mol^{-1}. This is much larger than 30 kJ mol^{-1}, and so we predict that $H_2(g)$ will be stable with respect to 2H(g).

Usually when comparing a molecule with free atoms, the enthalpy term is more important than the entropy term. Instead of having to consider thermodynamic cycles each time, therefore, we can make the following generalization:

Molecular orbital theory predicts that if the total energy of the electrons in a molecule is less (more negative) than the total electron energy of its constituent atoms, then that molecule will be stable with respect to the free atoms.

6.2.1 BONDING AND ANTIBONDING ORBITALS

So far we have only one molecular wavefunction, but for molecules with more than two electrons we are obviously going to need more. Let's start by having another look at how we combined the 1s orbitals to form the molecular orbital for H_2^+ and H_2. We considered the 1s orbital as drawn in Figure 40a and we combined it with another

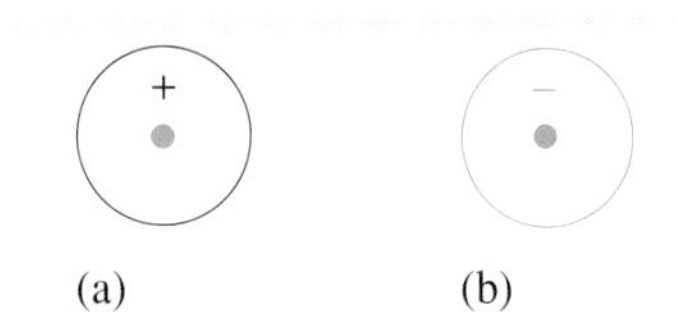

Figure 40 1s orbitals with (a) positive phase and (b) negative phase.

similar one. But what about the 1s orbital in Figure 40b? This is exactly the same as that in Figure 40a except for the phase.

Do we know what phase a 1s orbital has? In an atom, any property we want to calculate, such as the energy or the electron distribution, involves the square of the wavefunction and this tells us nothing about the *sign* of the wavefunction. So no observable property of an isolated atom will tell us the phase of the 1s orbital. When we want to combine two 1s orbitals, however, it matters very much whether the two have the same sign or not. In the previous Section we took two 1s orbitals with the same sign, but suppose we took two 1s orbitals of opposite sign.

■ Figure 41 shows a diatomic molecule in which 1s orbitals of opposite phase are combining to form a molecular orbital. What can you say about the probability of an electron being found between the two nuclei?

□ Between the nuclei, the positive and negative 1s orbitals overlap. When we add together wavefunctions of opposite phase, they cancel each other out. There is therefore very little chance of finding the electron in the region between the nuclei. Midway between the nuclei, there is no chance of finding the electron because there is a **nodal plane** at right-angles to the molecular axis, where the electron density is zero.

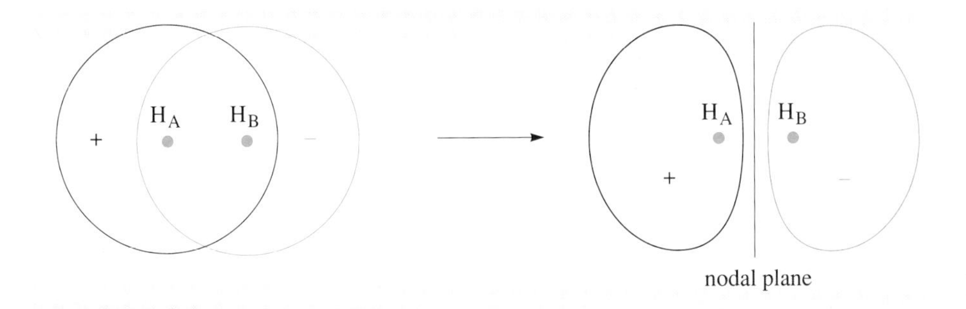

Figure 41 The combination of two 1s orbitals of opposite sign.

Figure 42 shows the probability of the electron being found at any point along the line joining the two nuclei for an electron in a 1s orbital, an electron in the orbital $N(1s_A + 1s_B)$ and an electron in the orbital we have just discussed. We shall label this second orbital $M(1s_A - 1s_B)$. The number in front, M, has changed because it is now the sum of $M^2(1s_A - 1s_B)^2$ over all space that must equal one. (It does not matter whether we take the combination $(1s_A - 1s_B)$ or the combination $(1s_B - 1s_A)$; both will have the same observable properties.)

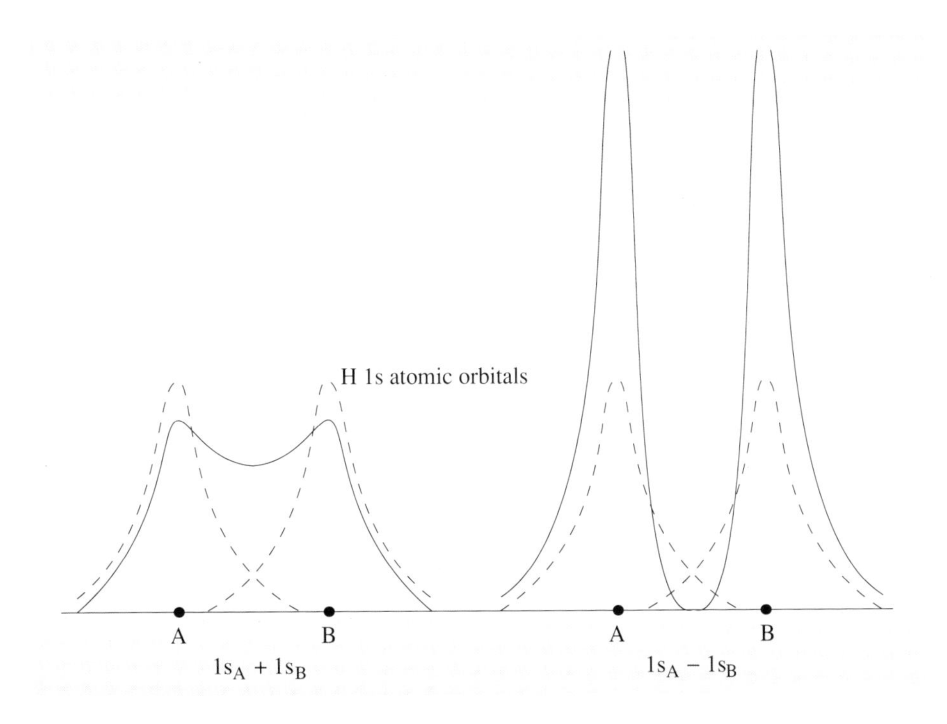

Figure 42 The probability of an electron being found at points along the line joining the two nuclei in dihydrogen for the combinations $N(1s_A + 1s_B)$ and $M(1s_A - 1s_B)$.

An electron in an $M(1s_A - 1s_B)$ orbital has even less chance of being attracted to both nuclei than one in a 1s orbital on one of the atoms. This orbital is therefore of *higher* energy than the 1s atomic orbital.

The orbital $N(1s_A + 1s_B)$ is known as a **bonding orbital**. An electron in a bonding orbital is very likely to be found between the nuclei, and thus draws the nuclei together or, in other words, bonds them together. Its energy will be less than that of the constituent atomic orbitals.

The orbital $M(1s_A - 1s_B)$ is an **antibonding orbital**. An electron in this orbital is less good at drawing the two nuclei together than an electron in an atomic orbital would be. In fact it tends to keep the nuclei apart. An electron in such an orbital has a higher energy than an electron in one of the atomic orbitals from which it is made.

Figure 43 shows a **molecular orbital energy-level diagram** for H_2. As with the energy-level diagram for H_2^+, energy increases from the bottom to the top of this diagram. Hence, the bonding orbital $N(1s_A + 1s_B)$ is placed below the atomic orbitals, and the antibonding orbital $M(1s_A - 1s_B)$ above. To remove an electron from the bonding orbital requires more energy than to remove one from the atomic 1s orbital of hydrogen.

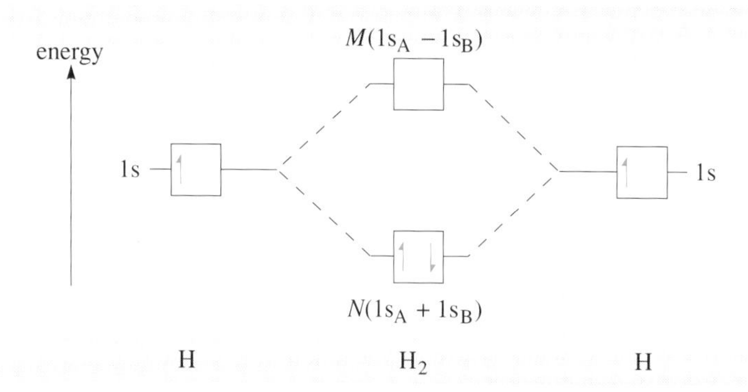

Figure 43 Molecular orbital energy-level diagram for H_2, showing the antibonding orbital.

SAQ 25 The molecule dihelium, He_2, would have four electrons. Use the orbital energy-level diagram for H_2 (Figure 43), and the rules that we used to assign electrons to orbitals in atoms, to determine the orbital energy-level diagram for this molecule. Represent the electrons as arrows in boxes. How will the energy of this molecule compare with that of two He atoms? He_2 has never been observed; does your energy-level diagram suggest a reason why this should be so?

6.2.2 MOLECULAR ORBITALS FROM s ORBITALS

We are now going to see how we can make molecular orbitals from atomic orbitals other than the 1s by applying the same principles that we used to construct the $N(1s_A + 1s_B)$ and $M(1s_A - 1s_B)$ molecular orbitals. Let's start with the 2s orbitals. A 2s orbital is spherically symmetric, like a 1s orbital, and we can combine it with another 2s orbital in exactly the same way as we combined two 1s orbitals. The factors N and M will differ because 2s wavefunctions are not exactly the same as 1s, but in writing combinations of orbitals from now on we shall omit these factors. Thus we shall represent $N(1s_A + 1s_B)$ by $(1s_A + 1s_B)$.

■ What molecular orbitals do you think we could make from two 2s orbitals, one on each of two nuclei A and B?

□ We could combine two 2s orbitals of the same phase to make $(2s_A + 2s_B)$, or we could combine two 2s orbitals of opposite phase to make $(2s_A - 2s_B)$.

Consider the molecule Li_2. Lithium, Li, has the electron configuration $1s^22s$, and Li_2 will therefore have six electrons. Suppose we combine two 1s atomic orbitals to make two molecular orbitals $(1s_A + 1s_B)$ and $(1s_A - 1s_B)$, and two 2s atomic orbitals to make two more. What energies will these orbitals have? The $(1s_A + 1s_B)$ molecular orbital will be of lower energy than a lithium 1s atomic orbital. The $(1s_A - 1s_B)$ molecular orbital will be of higher energy than the 1s atomic orbital. Similarly, the $(2s_A + 2s_B)$ molecular orbital will be of lower energy and the $(2s_A - 2s_B)$ molecular orbital of higher energy than the 2s orbital in the lithium atom. Will the $(2s_A + 2s_B)$ orbital be lower or higher in energy than the $(1s_A - 1s_B)$ orbital? This depends on whether the difference in energy between the 1s and 2s orbitals is greater than that between bonding and antibonding molecular orbitals from the same atomic orbitals. The energy difference between the 1s and 2s orbitals is large, 8×10^{-18} J for Li, and for all diatomic molecules the 1s antibonding molecular orbital is found to be of lower energy than the 2s bonding orbital. Each orbital can contain two electrons of opposite spin, so we can now feed our electrons into the Li_2 orbitals, using the same rules as we employed in the case of atoms. The first two go into $(1s_A + 1s_B)$, the next two into $(1s_A - 1s_B)$ and the last two into $(2s_A + 2s_B)$. The orbital energy-level diagram is shown in Figure 44.

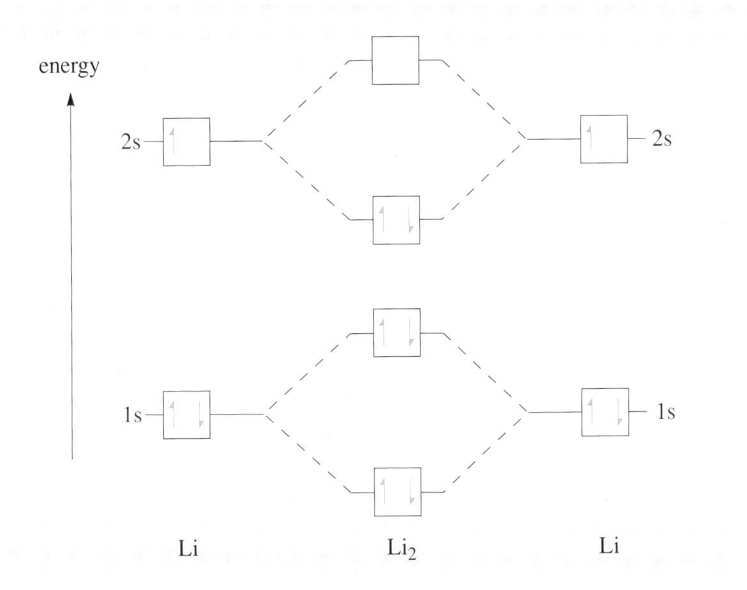

Figure 44 Orbital energy-level diagram for Li_2.

■ Would you expect Li_2 to be stable with respect to two Li atoms?

□ Yes, because the electrons in the $(2s_A + 2s_B)$ orbital are of lower energy than they would be in a 2s orbital. The energy contributions of the $(1s_A + 1s_B)$ and $(1s_A - 1s_B)$ electrons effectively cancel out relative to the 1s atomic orbital.

Li_2 illustrates one of the shortcomings of our theory so far. We have predicted that Li_2 is stable with respect to lithium atoms, so we might expect to be able to observe dilithium molecules. If we heat lithium up sufficiently, we do indeed find Li_2 molecules present in the vapour but, at normal temperatures and pressures, lithium is a solid metal in which a very large number of lithium atoms are bound together. We could not have predicted this because we have considered only the relative stabilities of Li_2 and Li. When we say that a particular molecule is stable with respect to the atoms from which it is made, you should bear in mind that it may not be stable with respect to other molecules or to solid forms of the element or compound.

Before we go on to discuss other molecules, there are some definitions and labels that need to be introduced.

Up till now we have labelled our molecular orbitals by using the labels of the atomic orbitals used to make them. This is fine for simple molecules but can become rather cumbersome for larger ones. (Imagine a label for an orbital in PF_5, formed by adding 2s orbitals on all six atoms!) These days, therefore, chemists label molecular orbitals according to their symmetry properties. The labels tell us how an orbital behaves when we act on it with the operations of the symmetry point group.

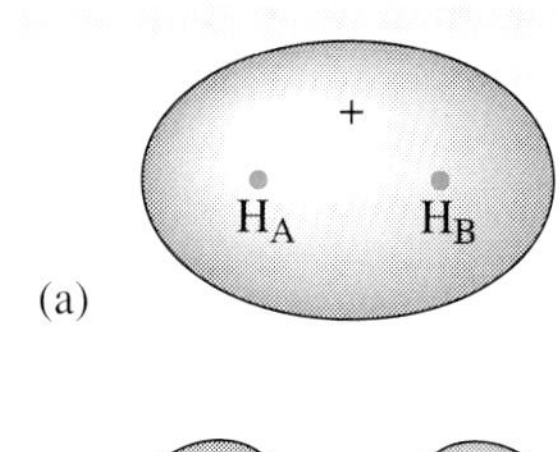

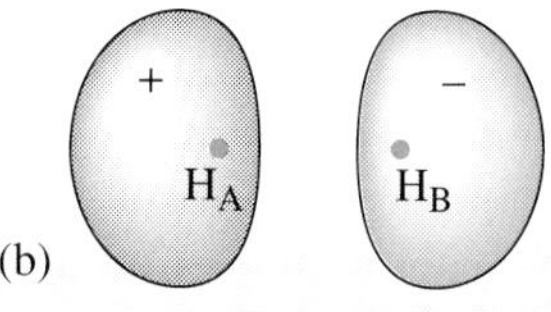

Figure 45 Boundary surfaces for (a) the $(1s_A + 1s_B)$ and (b) the $(1s_A - 1s_B)$ molecular orbitals.

For diatomic molecules we label the molecular orbitals according to how they behave when we rotate the molecule about the molecular axis (the C_∞ axis). If we do this to the $(1s_A + 1s_B)$ orbital, we find that whatever angle we rotate it through it always looks the same. Figure 45 shows **boundary surfaces** (the 95% contour in three dimensions) for $(1s_A + 1s_B)$ and $(1s_A - 1s_B)$. Orbitals with this symmetry are called **σ orbitals** (sigma, the Greek equivalent of s).

The orbital $(1s_A - 1s_B)$ will also give an identical orbital if we rotate it about the molecular axis. It too is a σ orbital. The two orbitals do not, however, have all the same symmetry properties. Look at Figure 45.

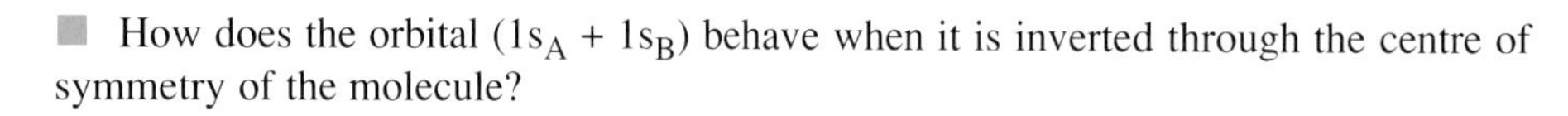

■ How does the orbital $(1s_A + 1s_B)$ behave when it is inverted through the centre of symmetry of the molecule?

□ It is turned into an identical orbital.

■ How does the orbital $(1s_A - 1s_B)$ behave when it is inverted through the centre of symmetry of the molecule?

□ Inversion of this orbital through the centre of symmetry of the molecule gives *minus* the original orbital (see Figure 46).

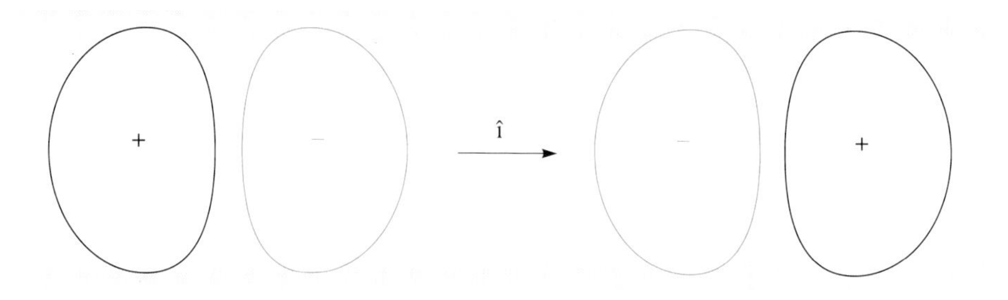

Figure 46 Inversion of the orbital $(1s_A - 1s_B)$ through the centre of symmetry of the molecule.

Orbitals like $(1s_A + 1s_B)$, which give an identical orbital if inverted through the centre of symmetry of the molecule are distinguished by the subscript g†.

Orbitals like $(1s_A - 1s_B)$, which give minus the original orbital when inverted through the centre of symmetry, are given the subscript u. So $(1s_A + 1s_B)$ is a σ_g orbital and $(1s_A - 1s_B)$ is a σ_u orbital.

■ How would you label $(2s_A + 2s_B)$ and $(2s_A - 2s_B)$?

□ $(2s_A + 2s_B)$ is σ_g, and $(2s_A - 2s_B)$ is σ_u.

We now have two σ_g orbitals and two σ_u orbitals. If we need to distinguish them we can do so by labelling them with the numbers 1, 2, 3, ... starting with the lowest energy orbital with a particular symmetry label. So $(1s_A + 1s_B)$ becomes $1\sigma_g$ and $(2s_A + 2s_B)$ becomes $2\sigma_g$. These numbers tell us nothing more than the order of energies of the molecular orbitals; *they are not quantum numbers*. Alternatively, we can label them with the atomic orbitals they are from, so that $(1s_A + 1s_B)$ becomes $1s\sigma_g$ and $(2s_A + 2s_B)$ becomes $2s\sigma_g$. Figure 47 shows the orbital energy-level diagram for Li_2 using these symmetry labels.

† g stands for the German *gerade*, meaning even; u stands for *ungerade*, meaning odd.

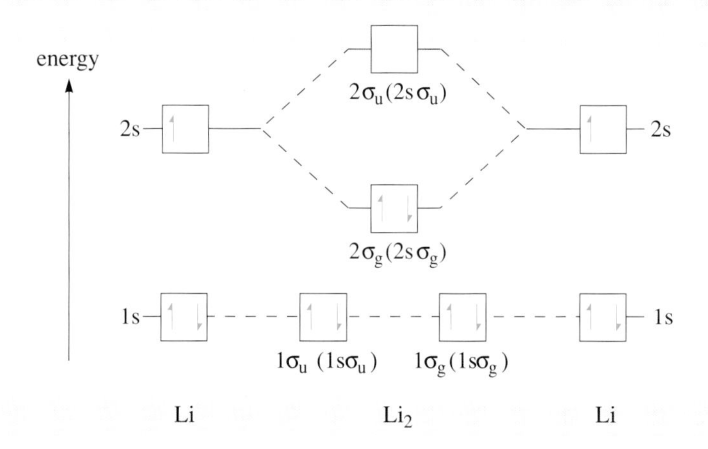

Figure 47 Orbital energy-level diagram for Li_2, showing symmetry labels.

You have probably noticed there is one further difference between Figures 44 and 47. In Figure 47 the $1\sigma_g$ and $1\sigma_u$ orbitals are drawn as having the same energy as the 1s orbitals. Why should this be so?

■ When we discussed atoms with more than one electron, we pointed out that, although the orbitals in these atoms are similar to those of the hydrogen atom, they were not identical with hydrogen atom orbitals. Can you remember what properties we mentioned in particular as varying from atom to atom?

□ The size and energy of the orbitals.

■ How do the sizes of the 1s orbitals in Li and H compare?

□ The 1s orbital in Li is smaller (Figure 35).

■ The distance between the hydrogen atoms in H_2 is 74 pm. The Li—Li bond length in Li_2 is 270 pm. How does the overlap of two 1s orbitals in H_2 compare with the overlap in Li_2?

□ The Li 1s orbitals are smaller than those of hydrogen and further apart. The overlap of the H 1s orbitals is much greater.

The difference in energy between the $1\sigma_g$ and 1s orbitals is a lot smaller in Li_2 than in H_2. This is illustrated in Figure 48. Remember that the electrons will be found within the contours shown in the diagram 95% of the time.

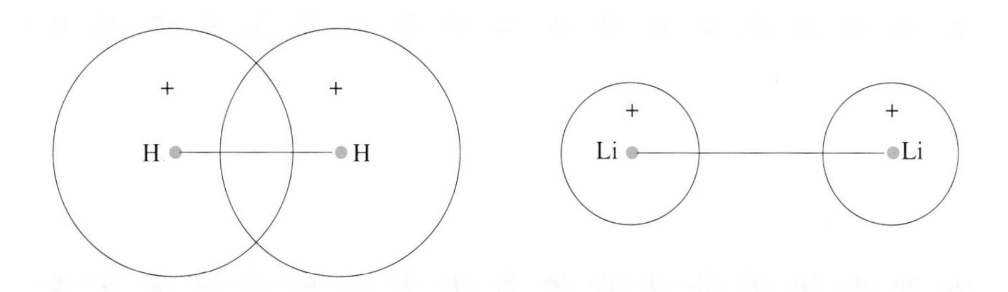

Figure 48 Overlap of 1s orbitals in dihydrogen and dilithium.

We find that, in general, there is very little overlap between atomic orbitals lower in energy than those of the valence-shell electrons. These orbitals are often left out of molecular orbital diagrams. This makes it easier to concentrate on the chemically more interesting valence orbitals.

Now try the following SAQ to see if you have understood how to form molecular orbitals from s orbitals. Then we go on to look at molecules where we need to consider p orbitals.

SAQ 26 Draw an orbital energy-level diagram for diberyllium, Be_2. Label the molecular orbitals with their symmetry labels and represent the electrons as arrows in boxes. A recent research report stated that the spectrum of Be_2 had been observed at low temperature. Would you have predicted the existence of Be_2 from your diagram? What is the normal form of beryllium at room temperature and atmospheric pressure?

6.2.3 MOLECULAR ORBITALS FROM p ORBITALS

So far, we have considered only molecular orbitals made by combining s atomic orbitals. We shall now see how we can combine 2p atomic orbitals. There are three 2p orbitals, $2p_x$, $2p_y$ and $2p_z$, directed along three perpendicular axes (Figure 49). Can we combine any one of these with any other one, or are there restrictions?

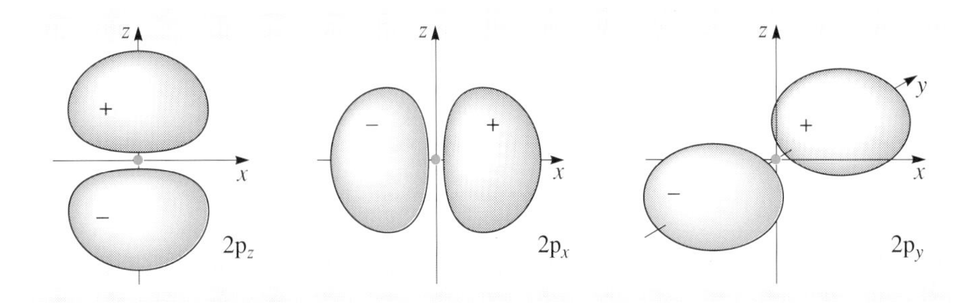

Figure 49 $2p_x$, $2p_y$ and $2p_z$ orbitals.

First we have to decide which direction is x, which is y and which is z, for the molecule. It is conventional to take the z direction along the axis of highest symmetry of a molecule.

■ What is the axis of highest symmetry for a diatomic molecule, such as H_2?

□ The molecular axis.

■ What order is this axis?

□ ∞. The molecular axis of diatomic and linear polyatomic molecules is always an axis of infinite order, C_∞.

Figure 50a shows a $2p_z$ orbital on nucleus A in the molecule N_2. A $2p_z$ orbital on nucleus B is shown in Figure 50b. Note that a $2p_z$ orbital is defined as having the positive lobe of the wavefunction in the positive z direction. This is merely a convention, because we do not know which way round the positive and negative lobes are. The convention is necessary to allow us to say whether a particular combination of 2p orbitals will overlap to increase the wavefunction between the nuclei or cancel it out.

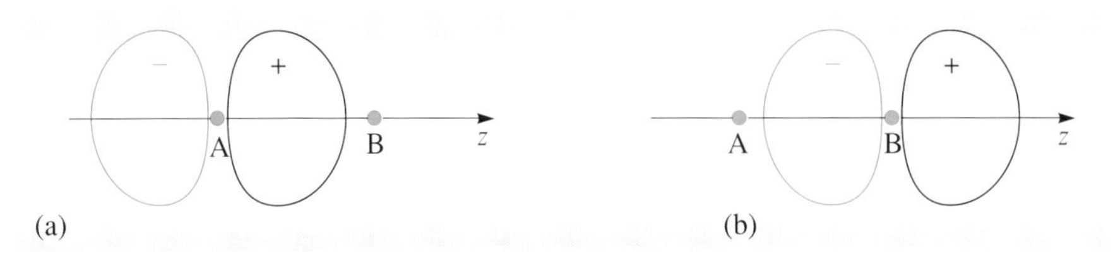

Figure 50 $2p_z$ orbitals on (a) nucleus A and (b) nucleus B in dinitrogen.

The x and y directions are at right-angles to the z direction and each other. We shall define the y direction as being in the plane of the paper and the x direction as coming up out of the paper. A $2p_x$ and a $2p_y$ orbital on atom A are shown in Figure 51.

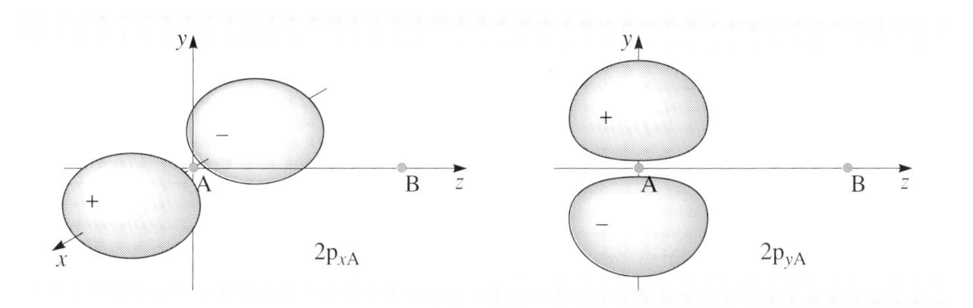

Figure 51 $2p_x$ and $2p_y$ orbitals on nucleus A of dinitrogen.

Let's start by combining two $2p_z$ wavefunctions, one on atom A and one on atom B.

■ Can you say anything about the combination of $2p_z$ orbitals as they are drawn in Figure 50?

□ If you look at this Figure you will see that, between the nuclei, one orbital has a positive lobe and the other a negative lobe. They will therefore cancel each other out and there will be little chance of finding the electron between the nuclei.

The molecular orbital $(2p_{zA} + 2p_{zB})$ will therefore be an antibonding orbital of higher energy than a $2p_z$ atomic orbital. Can we give this orbital a symmetry label?

■ Is the orbital $(2p_{zA} + 2p_{zB})$ a σ orbital?

□ Yes. In a linear molecule the label σ applies to any orbital that is unchanged by rotation about the molecular axis through any angle. Remember that diagrams such as Figure 50b are only cross-sections through a three-dimensional boundary surface. If you looked at a $2p_z$ orbital along the z axis, it would look like the $2p_x$ orbital shown in Colour Plate 2 at the back of the Block.

■ If we invert the orbital through the centre of symmetry of the molecule, do we obtain an identical orbital?

□ No. We produce minus the original orbital; $(2p_{zA} + 2p_{zB})$ must therefore take the subscript u.

So this orbital is a σ_u orbital. The $2p_z$ orbital in any atom other than hydrogen is higher in energy than the 1s and 2s orbitals. The σ_u orbital made from $2p_z$ orbitals is therefore higher in energy than the σ_u orbitals made from the 1s and 2s orbitals. The full label for $(2p_{zA} + 2p_{zB})$ is either $3\sigma_u$ because it is the third lowest energy σ_u orbital, or $2p\sigma_u$ because it is made from 2p orbitals.

■ What combination of $2p_z$ orbitals will give us a bonding orbital?

□ We can make a bonding combination by reversing the signs on one of the $2p_z$ atomic orbitals. This gives us the molecular orbital $(2p_{zA} - 2p_{zB})$ shown in Figure 52.

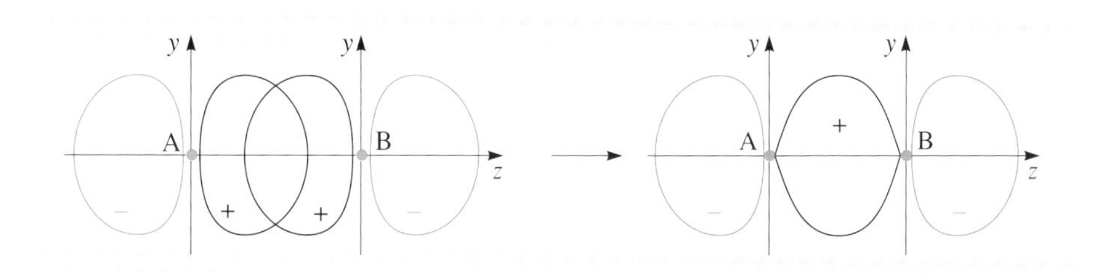

Figure 52 A bonding molecular orbital made by combining two $2p_z$ atomic orbitals.

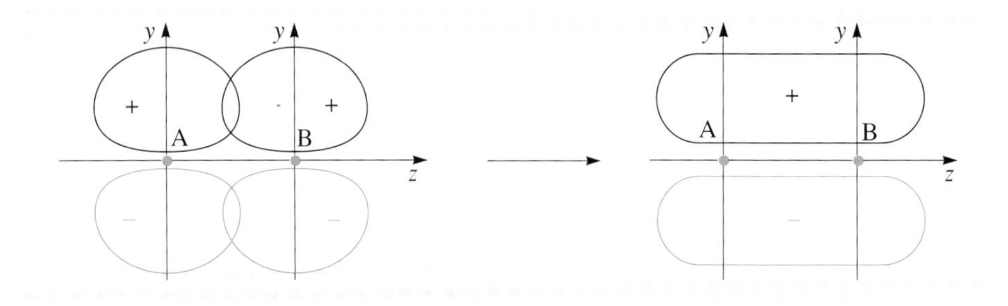

Figure 53 The bonding orbital ($2p_{yA} + 2p_{yB}$).

This orbital will be labelled $3\sigma_g$ or $2p\sigma_g$. Although it looks different from $1\sigma_g$ and $2\sigma_g$ because of the change of sign and the zero probability at the nuclei, it is still labelled σ_g since it behaves in the same way as $1\sigma_g$ and $2\sigma_g$ when rotated about the molecular axis or inverted through the centre of symmetry.

Now let's combine two $2p_y$ orbitals. Figure 53 shows the combination ($2p_{yA} + 2p_{yB}$). In this combination the amplitude of the wavefunction is increased between the nuclei but only above and below the molecular axis. The probability of finding the electron on the molecular axis in this orbital is zero; that is, the molecular axis is a nodal plane. The high probability just above and below this line, however, does serve to draw the nuclei together, and so this orbital is bonding. It is a different sort of bonding orbital from the σ_g orbitals we have already met. ($2p_{yA} + 2p_{yB}$) produces an identical orbital only if we rotate it about the molecular axis through a complete revolution (2π radians). If we rotate it through half a revolution (π radians) about this axis, we produce minus the original orbital. Orbitals like ($2p_{yA} + 2p_{yB}$) are called **π orbitals** (pi, the Greek equivalent of p).

■ Should ($2p_{yA} + 2p_{yB}$) be given the subscript g or u?

□ Inversion through the centre of symmetry produces minus the original orbital, so the correct subscript is u.

So ($2p_{yA} + 2p_{yB}$) is a bonding orbital, to which we give the symbol π_u.

■ What would the antibonding orbital formed from $2p_y$ orbitals be?

□ ($2p_{yA} - 2p_{yB}$). This has the symbol π_g, and is shown in Figure 54.

We can make, therefore, two molecular orbitals, one bonding (π_u) and one antibonding (π_g), by combining $2p_y$ atomic orbitals. Can we make two molecular orbitals by combining $2p_x$ orbitals? Yes. The $2p_x$ orbitals can form two combinations. These will be identical with those of $2p_y$, except that they will be concentrated in different directions.

Atomic orbitals that differ only in direction (for example, $2p_x$, $2p_y$ and $2p_z$) are **degenerate**; that is, the electrons in them have the same energy. Similarly, the molecular orbitals ($2p_{xA} + 2p_{xB}$) and ($2p_{yA} + 2p_{yB}$) will be degenerate. The two degenerate bonding orbitals are given the symbol $1\pi_u$ or $2p\pi_u$. The two degenerate antibonding orbitals are given the symbol $1\pi_g$ or $2p\pi_g$.

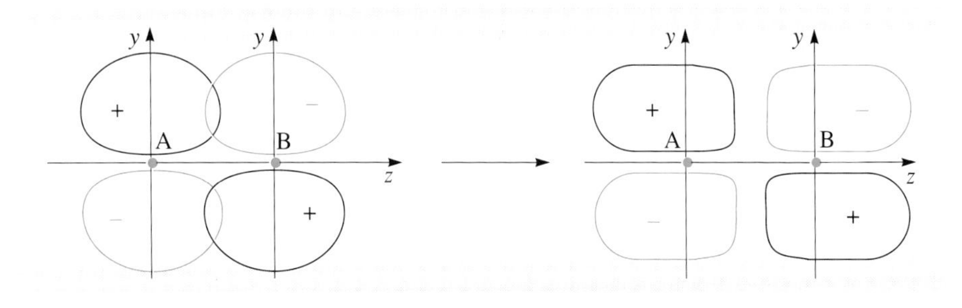

Figure 54 An antibonding π orbital.

σ orbitals, like s orbitals, are non-degenerate, whereas there are two π orbitals of the same energy.

Let us try, then, to construct an orbital energy-level diagram for a molecule that has molecular orbitals made from 2p orbitals – the dinitrogen molecule, N_2. The nitrogen atom has seven electrons, so N_2 will have fourteen. Which molecular orbitals shall we put them into? We always have to start with those of lowest energy, so we begin with those formed from the nitrogen 1s orbitals. We put two electrons with paired spin in $1s\sigma_g$ and two in $1s\sigma_u$. Then we can put two each in the molecular orbitals formed from the 2s atomic orbitals, the $2s\sigma_g$ and the $2s\sigma_u$. This leaves six electrons.

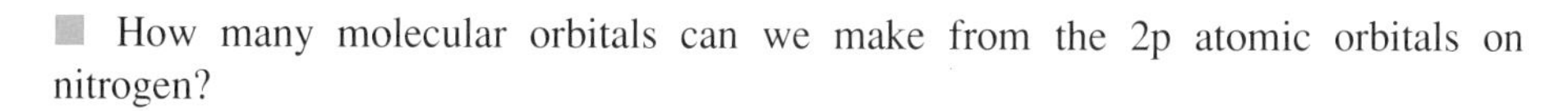

■ How many molecular orbitals can we make from the 2p atomic orbitals on nitrogen?

□ Six. $2p\sigma_g$, $2p\sigma_u$, $2p\pi_u$ (two orbitals) and $2p\pi_g$ (two orbitals).

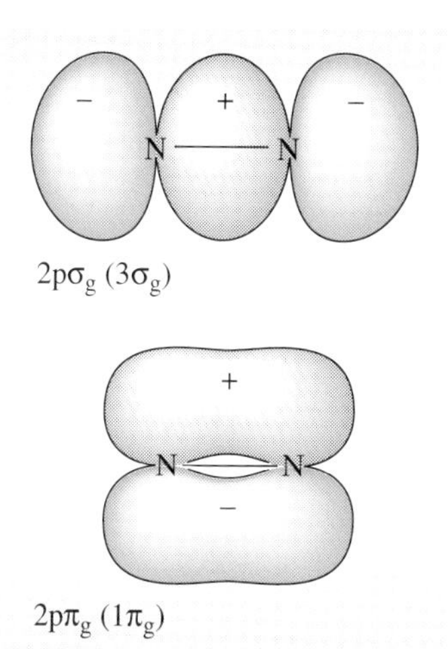

Figure 55 $2p\sigma_g$ and $2p\pi_u$ orbitals in dinitrogen.

$2p\sigma_g$ and $2p\pi_u$ were the bonding combinations, so they must be of lower energy than the 2p orbital and will be the next ones to be filled. Which one will be filled first? Figure 55 reproduces the $2p\sigma_g$ and one of the $2p\pi_u$ orbitals. In the $2p\sigma_g$ orbital, nuclei are held together by electron density directly between them, whereas in $2p\pi_u$ the electron density is holding them together above and below the molecular axis. It seems reasonable that an electron in $2p\sigma_g$ would be better at drawing the nuclei together than one in $2p\pi_u$, and theoretical calculations show that for orbitals made purely from p orbitals this is so.

So we fill the $2p\sigma_g$ first and then the $2p\pi_u$. The $2p\sigma_g$ can take two electrons, and the $2p\pi_u$ four, and this disposes of all the remaining electrons. Figure 56 shows the orbital energy-level diagram for dinitrogen.

■ How many electrons in bonding orbitals are there in N_2?

□ Ten: those in $1s\sigma_g$, $2s\sigma_g$, $2p\sigma_g$ and $2p\pi_u$.

■ How many electrons are there in antibonding orbitals?

□ Four: those in $1s\sigma_u$ and $2s\sigma_u$.

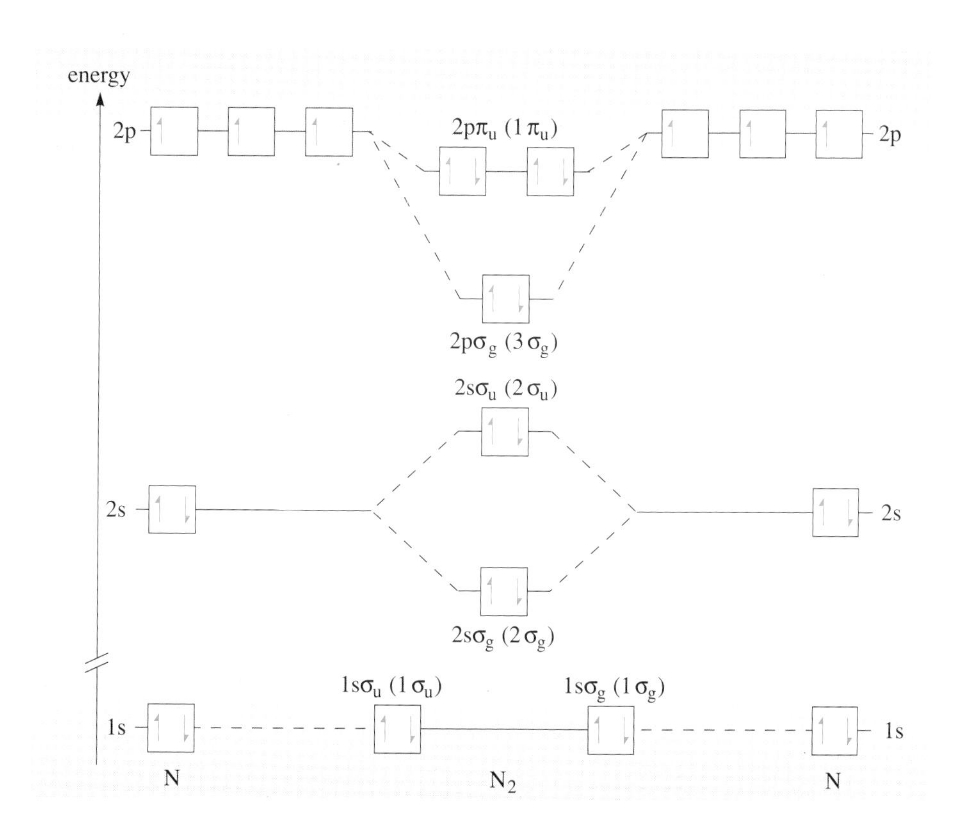

Figure 56 Orbital energy-level diagram for dinitrogen.

Dinitrogen has an excess of six electrons in bonding orbitals over those in antibonding orbitals. The structural formula of dinitrogen, N≡N, shows the two atoms joined by three pairs of electrons in covalent bonds. In molecular orbital theory the number of bonds between two atoms is determined by the excess number of pairs of electrons in bonding orbitals over those in antibonding orbitals.

We can define a quantity called the **bond order**, which is equal to the number of pairs of electrons in bonding orbitals minus the number of pairs in antibonding orbitals. An unpaired electron counts as half a pair and thus contributes $+\frac{1}{2}$ if it is in a bonding orbital or $-\frac{1}{2}$ if it is in an antibonding orbital. In general, we find that the bond order corresponds to the number of electron pairs covalently bonding the atoms together in the Lewis structure.

■ What is the bond order of dinitrogen?

□ Three. There are five pairs of electrons in bonding orbitals and two in antibonding orbitals.

In the next Section we are going to look at some more diatomic molecules.

SUMMARY OF SECTION 6.2.3

1 From six 2p atomic orbitals (three on each atom) we can make six molecular orbitals.

2 Each pair of 2p orbitals will combine to form two molecular orbitals, one bonding and one antibonding.

3 The two orbitals formed from the $2p_z$ atomic orbitals are unchanged by rotation through any angle about the molecular axis and are therefore σ orbitals.

4 The $2p_x$ and $2p_y$ orbitals form π orbitals. The sign of these orbitals is reversed when they are rotated through half a revolution about the molecular axis.

5 σ bonding orbitals are unchanged by inversion through the centre of symmetry of the molecule and are therefore labelled σ_g. π bonding orbitals are changed into minus themselves and are labelled π_u.

6 σ antibonding orbitals are labelled σ_u and π antibonding orbitals π_g.

7 σ orbitals are non-degenerate; π orbitals are doubly degenerate (that is, there are always two of the same energy).

8 The σ bonding orbital is of lower energy than the π bonding orbitals.

SAQ 27 What are the bond orders of the following molecules: (a) H_2; (b) He_2; (c) Li_2?

7 DIATOMIC MOLECULES

7.1 HOMONUCLEAR MOLECULES OF THE FIRST AND SECOND ROWS OF THE PERIODIC TABLE

A **homonuclear diatomic molecule** is one in which both nuclei are the same, for example H_2 and N_2. In the first row of the Periodic Table, H_2 is the only example. From the second row we have N_2, O_2 and F_2, which are stable under normal conditions of temperature and pressure. We looked at N_2 in the previous Section. Here we shall consider the molecular orbital description of O_2, and use it as an example of how we can use the theory to explain and/or predict properties of molecules.

As a further example we shall then take C_2. This is not a stable molecule under normal conditions, but it has been detected in such places as flames and the heads of comets.

Oxygen has the electron configuration $1s^22s^22p^4$, and so the dioxygen molecule, O_2, has sixteen electrons. Four of these will go into the two σ orbitals formed from 1s orbitals, and four more into the two σ orbitals formed from 2s orbitals. This leaves eight electrons to go into molecular orbitals formed by combining 2p orbitals. The orbitals available will be similar to those for N_2 but will have different energies. So we can put two electrons into the σ bonding orbital ($3\sigma_g$ or $2p\sigma_g$) and two into each of the π bonding orbitals ($1\pi_u$ or $2p\pi_u$). This disposes of six of the eight electrons. What do we do with the last two?

■ Would you expect these two to go into the π antibonding orbital, $2p\pi_g$, or the σ antibonding orbital, $2p\sigma_u$?

□ They will go into whichever has the lower energy. When we form a bonding and an antibonding orbital from two atomic orbitals, the antibonding orbital is raised in energy by about the same amount that the bonding orbital is lowered in energy. Since the $2p\sigma_g$ bonding orbital is lower in energy than the $2p\pi_u$ bonding orbitals, the $2p\sigma_u$ antibonding orbital will be *higher* in energy than the $2p\pi_g$ antibonding orbitals. The electrons will therefore go into the $2p\pi_g$ ($1\pi_g$) orbitals. The orbital energy-level diagram for O_2 is shown in Figure 57.

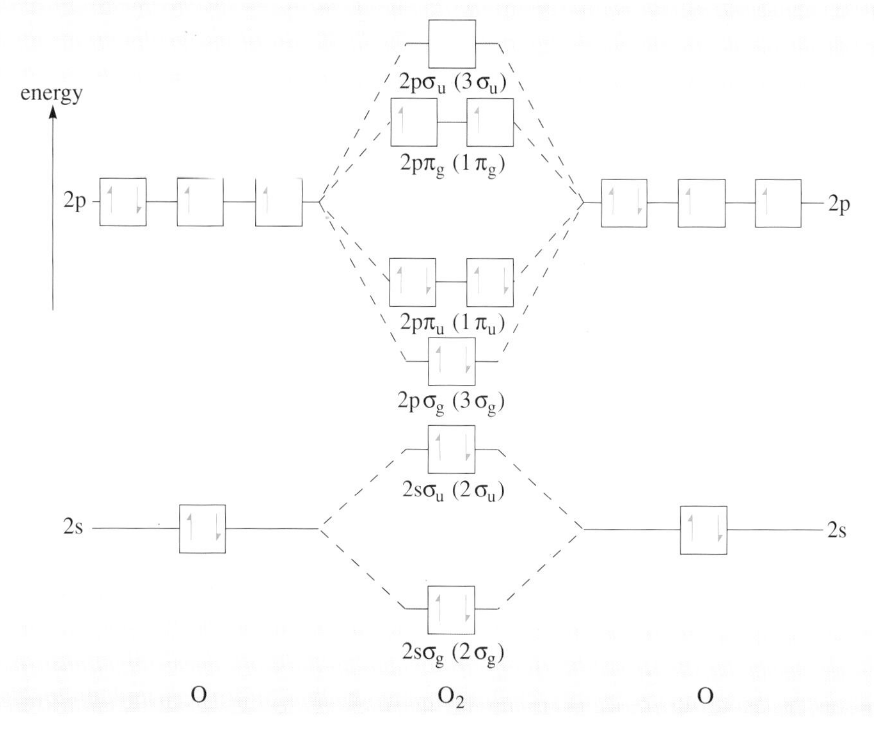

Figure 57 Orbital energy-level diagram for dioxygen.

■ In Figure 57 we have shown the two electrons in the $1\pi_g$ orbitals with parallel spins. Why have we done this?

□ Electrons in molecules as well as those in atoms obey Hund's rule (Section 6.1.1), and since the two $1\pi_g$ orbitals are degenerate the two electrons will go one into each $1\pi_g$ orbital with parallel spins.

As you saw in the video sequence 7, if we place liquid oxygen (which consists of O_2 molecules) near a strong magnet, it is drawn into the magnetic field. Liquid nitrogen, on the other hand, is hardly affected and is if anything slightly repelled from the field. Dioxygen is said to be **paramagnetic**, whereas dinitrogen is said to be **diamagnetic**. Atoms or molecules with unpaired spins have a magnetic moment and interact with an applied magnetic field. By predicting that O_2 has two unpaired electrons, the molecular orbital theory can account for the behaviour of dioxygen in a magnetic field. It also predicts that dinitrogen will be diamagnetic because all the electrons are paired.

The explanation of the paramagnetism of dioxygen was one of the early triumphs of molecular orbital theory, since the property could not be explained in terms of earlier theories.

■ What is the bond order in O_2?

□ Two: O_2 has three pairs of electrons in 2p bonding orbitals and two half-pairs in antibonding orbitals.

As with N_2, the bond order is equal to the number of covalent bonds predicted by the Lewis structure. Are the different bond orders of O_2 (two) and N_2 (three) reflected in their properties?

We would expect that the more electrons there are holding two nuclei together, the harder it will be to force them apart and the closer they will be at equilibrium. Let's then compare the dissociation energies and bond lengths of N_2 and O_2. The dissociation energies, that is, $\Delta H^{\ominus}_{m}$ for the processes $N_2(g) \longrightarrow 2N(g)$ and $O_2(g) \longrightarrow 2O(g)$, are 945 kJ mol^{-1} for dinitrogen and 498 kJ mol^{-1} for dioxygen. It thus requires more energy to split the dinitrogen molecule into two nitrogen atoms than to obtain two oxygen atoms from O_2. This is what we would expect from the bond orders. Similarly, the bond length of N_2 (110 pm) is less than that of O_2 (121 pm).

When comparing bond lengths and dissociation energies in this way, you should be careful to choose molecules with similar molecular orbitals. In this case we are comparing two molecules that use the same atomic orbitals (2s and 2p) to form their molecular orbitals. We can thus compare N_2 with O_2 and F_2, but not with H_2 or Cl_2, if we are interested in the effect of bond order. The dissociation energies of H_2 and F_2, which both have bond orders of 1, are 436 kJ mol^{-1} and 158 kJ mol^{-1}, respectively.

7.1.1 THE MOLECULE C_2

■ Look again at the orbital energy-level diagrams for N_2 and O_2 (Figures 56 and 57). Which molecular orbitals would you expect to be filled in C_2?

□ C_2 has twelve electrons, so we would expect to fill the $1\sigma_g$, $1\sigma_u$, $2\sigma_g$, $2\sigma_u$, and $3\sigma_g$ orbitals, and have two electrons in the $1\pi_u$ orbitals.

Spectroscopic observations of C_2 indicate that in the ground state the molecule has no unpaired electrons.

■ Does this evidence fit the orbital occupancy we predicted above?

□ No. By Hund's rule the two electrons in the $1\pi_u$ orbitals would be expected to be unpaired.

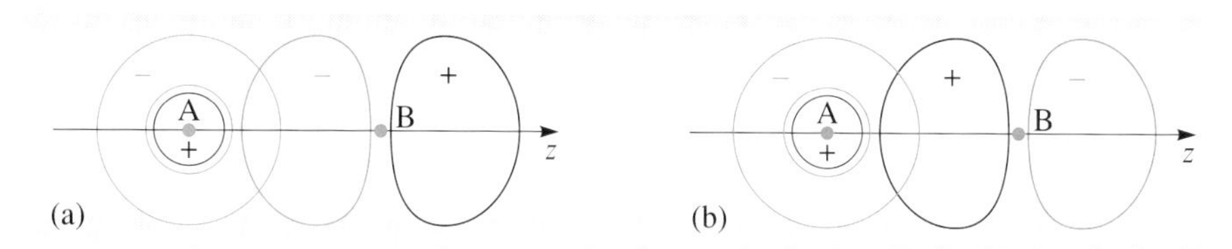

Figure 58 Overlap of a 2s orbital and a $2p_z$ orbital to form (a) the combination $(2s_A + 2p_{zB})$ and (b) the combination $(2s_A - 2p_{zB})$.

So far, we have made our molecular orbitals either entirely from s orbitals or entirely from p orbitals. This was because we can combine atomic orbitals only of roughly the same energy. For the hydrogen atom, however, 2s and 2p orbitals are of the same energy, and for the lighter atoms of the second row of the Periodic Table the difference between the energies of the 2s and 2p orbitals is still quite small. Perhaps, then, we should consider the overlap of 2s and 2p orbitals (Figure 58).

■ Will the combination shown in Figure 58a be bonding or antibonding?

□ Bonding. Between the nuclei the two wavefunctions have the same sign.

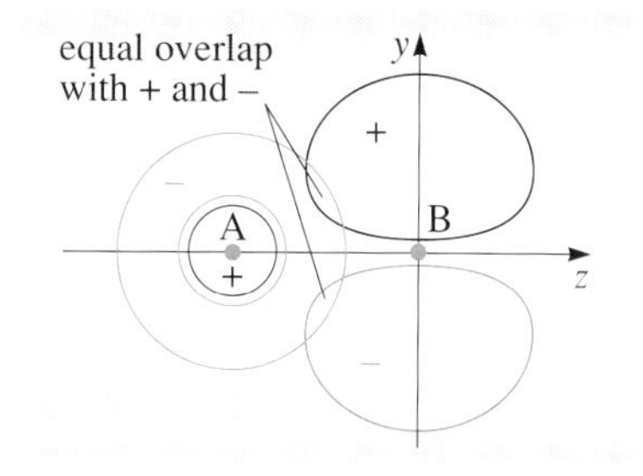

Figure 59 A $2s_A$ orbital and a $2p_{yB}$ orbital.

The combination shown in Figure 58b will be antibonding, because the 2s orbital will overlap with the positive lobe of the $2p_z$ orbital, thereby reducing the electron density between the nuclei.

In the combination of $2s_A$ and $2p_{yB}$ shown in Figure 59, the 2s orbital overlaps equally with both the positive and negative lobes of the $2p_y$ orbital, and the two overlaps will cancel out.

We cannot combine 2s with $2p_y$. Similarly, we cannot combine 2s and $2p_x$. It is possible to combine 2s and $2p_z$, which both form σ orbitals, but not 2s and $2p_x$ or $2p_y$; the latter two form π orbitals. This is an example of a rule that we shall meet later on when we study larger molecules, which says that only atomic orbitals with the same symmetry properties can be combined.

Does the possibility of overlap between 2s and $2p_z$ atomic orbitals affect the orbital energy-level diagram for C_2? If the carbon 2s and 2p atomic orbitals are close enough together in energy, then in constructing the $2\sigma_g$, $2\sigma_u$, $3\sigma_g$ and $3\sigma_u$ molecular orbitals we should consider the 2s and $2p_z$ atomic orbitals together. Let's first look at the formation of the $2\sigma_g$ and the $3\sigma_g$ molecular orbitals. We have regarded the $2\sigma_g$ orbitals so far as a combination of 2s orbitals, and the $3\sigma_g$ as a combination of 2p orbitals. In molecules like C_2, however, $2\sigma_g$ and $3\sigma_g$ will both be formed from a combination of 2s and 2p orbitals. Thus, $2\sigma_g$ will be mainly 2s orbitals, but with some contribution from the $2p_z$, whereas $3\sigma_g$ will consist mainly of $2p_z$ with a small contribution from 2s. The addition of $2p_z$ to $2\sigma_g$ will lower its energy as it will become more bonding. The addition of 2s to $3\sigma_g$, however, raises the energy of this orbital because it adds some antibonding character.

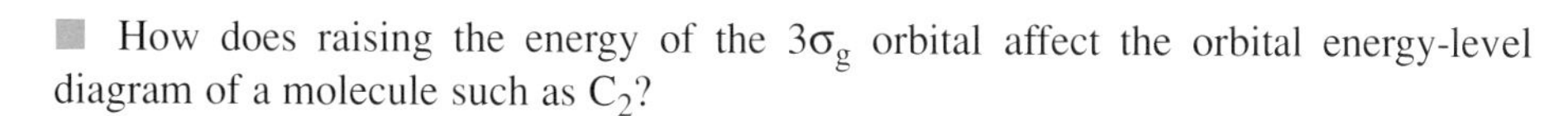

■ How does raising the energy of the $3\sigma_g$ orbital affect the orbital energy-level diagram of a molecule such as C_2?

□ If the energy of the $3\sigma_g$ orbital is raised sufficiently, it may become of higher energy than the $1\pi_u$.

■ If the $3\sigma_g$ orbital were higher in energy than the $1\pi_u$, would this explain why C_2 has no unpaired electrons in its ground state?

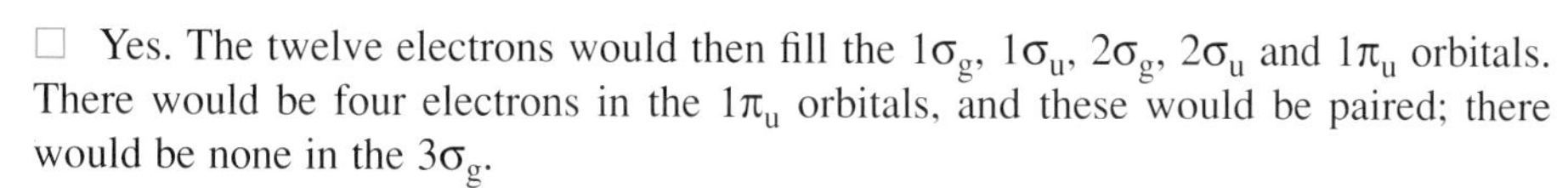

□ Yes. The twelve electrons would then fill the $1\sigma_g$, $1\sigma_u$, $2\sigma_g$, $2\sigma_u$ and $1\pi_u$ orbitals. There would be four electrons in the $1\pi_u$ orbitals, and these would be paired; there would be none in the $3\sigma_g$.

The revised orbital energy-level diagram for C_2 is shown in Figure 60.

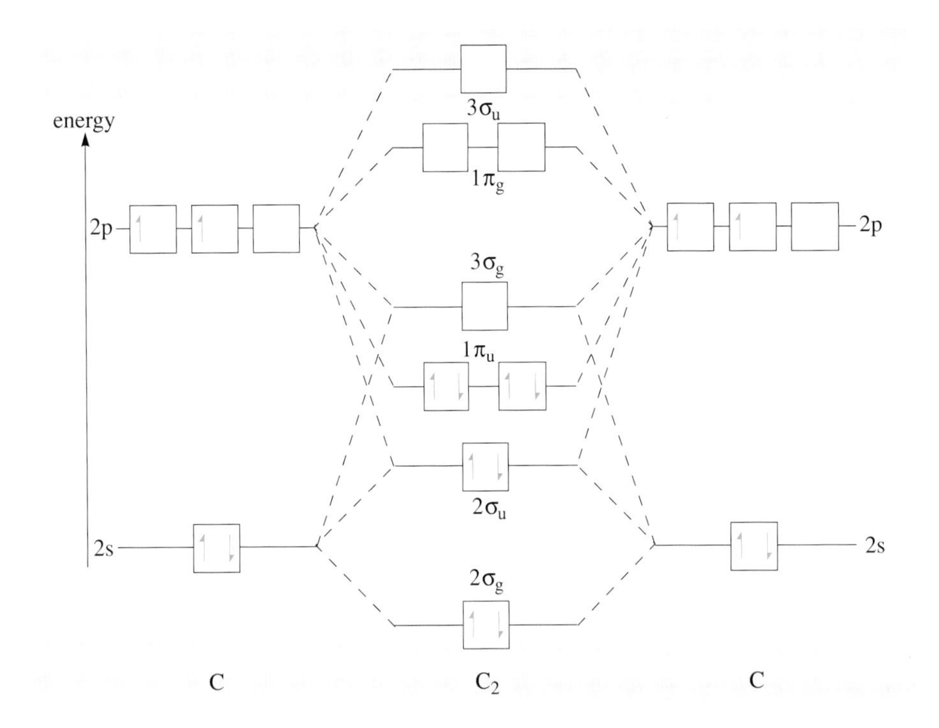

Figure 60 Orbital energy-level diagram for C_2.

How do we know when to use the order of orbital energies shown in Figure 60 and when to use the one that we used for N_2 and O_2? The 2s and 2p energy levels are the same for hydrogen, but gradually get further apart as we go across row 2 of the Periodic Table. By the time we get to fluorine, the 2s and 2p levels are sufficiently far apart for $2\sigma_g$ to be almost entirely 2s and $3\sigma_g$ to be almost entirely 2p. In between, $2\sigma_g$ will be mostly 2s orbitals with some contribution from 2p, and $3\sigma_g$ will be mainly 2p with some contributions from 2s. The changeover from the diagram in Figure 60 to that in Figure 57 comes between C_2 and O_2*.

A rough guide, which will also be useful for larger molecules, is that 2s and 2p should be considered together for the elements up to carbon and separately from oxygen onwards in the second row. For later rows, *n*s and *n*p can be considered together if the energy gap between them is less than or roughly equal to the 2s/2p gap in nitrogen.

7.2 SUMMARY OF SECTION 7.1

1 Molecular orbital theory describes the bonding in homonuclear diatomic molecules.

2 Molecular orbital occupancies in these diatomic molecules predict bond orders and magnetic properties. High bond orders correspond to large dissociation energies and short bond lengths.

3 If the 2s and 2p atomic orbitals are close in energy, then the molecular orbitals $2\sigma_g$ and $3\sigma_g$ are no longer purely 2s or 2p but mixtures of these. This can result in the $3\sigma_g$ orbital being of higher energy than the $1\pi_u$.

SAQ 28 Draw an orbital energy-level diagram for dineon, Ne_2. What is the bond order of this molecule? Would you predict *from your diagram* that this molecule should exist?

SAQ 29 Draw an orbital energy-level diagram for O_2^+. (Use the diagram for O_2 but feed in one fewer electron). How would you expect (a) the bond length and (b) the dissociation energy of this ion to differ from those of O_2?

SAQ 30 Use the orbital energy-level diagram for N_2 (Figure 56) to find the bond orders of N_2^+ and N_2^-. How would you expect the dissociation energies of these ions to compare with each other and with that of N_2?

* Very accurate calculations show that in N_2, the $3\sigma_g$ and $1\pi_u$ orbitals are very close in energy with the $1\pi_u$ just lower, showing that there is some interaction between 2s and 2p in N_2.

7.3 HOMONUCLEAR DIATOMIC MOLECULES OF LATER ROWS

Similar diagrams to those for N_2, O_2 and C_2 can be drawn for molecules composed of atoms from later rows. Take, for example, the chlorine molecule Cl_2. Chlorine has the electron configuration $1s^22s^22p^63s^23p^5$. Here we can regard the 1s, 2s and 2p orbitals as being effectively atomic orbitals, and consider only the valence shell orbitals 3s and 3p. In chlorine there is a relatively large energy gap between 3s and 3p and so, as in N_2 and O_2, we concentrate on the p orbitals. 3p orbitals combine in just the same way as 2p to form σ_g and π_u bonding orbitals and σ_u and π_g antibonding orbitals. If we number these orbitals, however, we have to allow for the 2s and 2p. So, the 1s orbitals would provide $1\sigma_g$ and $1\sigma_u$ molecular orbitals (these are labelled as molecular orbitals even though there is little overlap between 1s orbitals on the two atoms). The 2s would give $2\sigma_g$ and $2\sigma_u$, the 2p would give $3\sigma_g$, $1\pi_u$, $3\sigma_u$ and $1\pi_g$, and the 3s would give $4\sigma_g$ and $4\sigma_u$. The orbitals formed from the 3p orbitals are thus labelled $5\sigma_g$ (or $3p\sigma_g$), $2\pi_u$ (or $3p\pi_u$), $2\pi_g$ (or $3p\pi_g$) and $5\sigma_u$ (or $3p\sigma_u$). An orbital energy-level diagram for Cl_2 is shown in Figure 61.

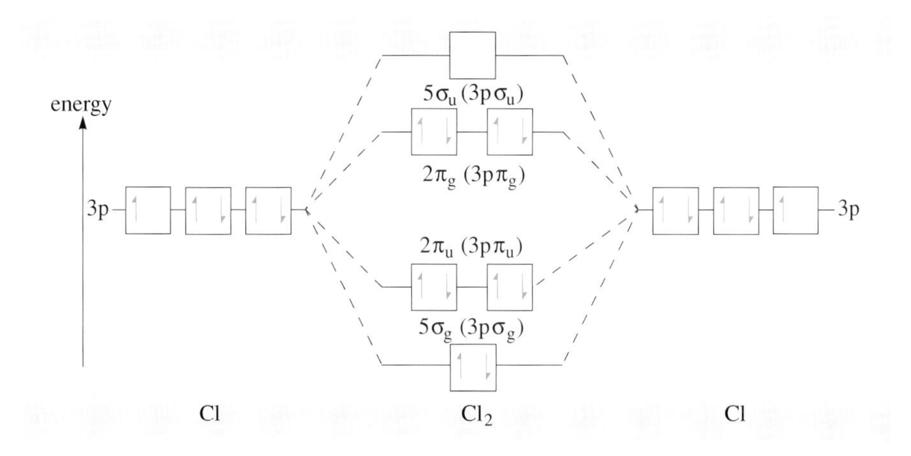

Figure 61 Orbital energy-level diagram for Cl_2.

If we went to Br_2, the 4p orbitals would form a similar set but the numbering would go even higher as we would have to allow for 1s, 2s, 2p, 3s, 3p, 4s and 3d orbitals. The labels $4p\sigma_g$ etc. are easier to use, because forgetting to count some of the lower energy orbitals becomes more likely.

SAQ 31 Although elemental sulphur is a solid under normal conditions of temperature and pressure, S_2 molecules can be observed in sulphur vapour when the solid is heated. Draw an orbital energy-level diagram for S_2, showing the orbitals made from the 3p on S. Predict the bond order of S_2 and state whether or not it will be paramagnetic.

7.4 HETERONUCLEAR DIATOMIC MOLECULES

Heteronuclear diatomic molecules that are formed from elements of the first and second rows of the Periodic Table and are found at normal temperatures and pressures are HF, CO and NO. Others that have been observed at high temperatures, in discharge lamps, in flames or in space are LiH, LiF, OH, BeH, BeO, BF, BH, CH, CN, and NH. Some of the molecules on this second list will be stable with respect to the two separate atoms but not with respect to other molecules or other forms of the compound. LiH, LiF and BeO are normally found as ionic solids. The other molecules are unstable with respect to covalent compounds in which the atoms have their normal valencies: H_2O, $(BeH_2)_n$, BF_3, B_2H_6, CH_4, $(CN)_2$ and NH_3.

As an example, let's look at the molecular orbitals of nitric oxide (NO). How do they differ from those of N_2 and O_2, and does molecular orbital theory enable us to predict the properties of NO? Nitric oxide at room temperature and atmospheric pressure is a colourless gas. The gas consists mainly of paramagnetic NO molecules, but at lower temperatures nitric oxide condenses to a liquid and then a solid containing dimers, N_2O_2. On contact with air, NO reacts with oxygen to form nitrogen dioxide, NO_2. It also reacts readily with halogens, other than iodine, to give nitrosyl halides, NOF, NOCl, and NOBr.

The NO molecule has fifteen electrons. This is an odd number and hence there must be at least one unpaired electron. The unpaired electron will account for the paramagnetism of the molecule, whichever theory of bonding we use. We would expect molecular orbital theory to explain why NO is stable with respect to N and O atoms, and to predict how the bond length and dissociation energy compare with those of N_2 and O_2.

To construct molecular orbitals for NO we first have to decide which atomic orbitals to combine. We can combine only atomic orbitals that are close in energy, so let's see what the atomic energy levels of N and O are. These can be found by looking up the ionization energies of electrons from the orbitals of interest. For nitrogen, the ionization energies are: 1s, 646×10^{-19} J; 2s, 32.6×10^{-19} J; 2p, 23.3×10^{-19} J; for oxygen the values are: 1s, 862×10^{-19} J; 2s, 45.6×10^{-19} J; 2p, 21.8×10^{-19} J. These are the energies required to *remove* the electrons, that is we have to *add* this amount of energy in order to free the electron from the atom. The electron thus gains this amount by being in the atom, and so its energy in the atom is minus this amount relative to a free electron. Thus we can take the values of the orbital energies to be minus the values of the ionization energies. The orbitals closest in energy are the 2p orbitals on nitrogen and the 2p orbitals on oxygen. The 1s orbitals are too far from each other and from any other orbitals to overlap significantly, so we can forget these. The energies of the 2s and 2p orbitals on each atom are quite well separated, so we shall concentrate only on the 2p as we did with N_2 and O_2.

The three 2p orbitals from each atom can be combined to give σ bonding and antibonding orbitals and π bonding and antibonding orbitals.

■ Will the bonding orbital be labelled σ_g or σ_u?

□ Neither label is appropriate. The subscripts g and u refer to behaviour under inversion through the centre of symmetry of the molecule. Because the two atoms are not identical, the molecule NO does not have a centre of symmetry.

For heteronuclear diatomic molecules, both σ bonding and the σ antibonding orbitals have the same symmetry and thus have to have the same label, σ. Where bonding and antibonding orbitals have the same symmetry label, it is common to distinguish between them by labelling the antibonding orbital with an asterisk. Thus the 2pσ antibonding orbital may be written 2pσ^* (two p sigma star).

We shall now construct a partial orbital energy-level diagram for NO using 2p orbitals on both N and O. Look at Figure 62: on one side of the diagram we have the N 2p orbitals and on the other side the O 2p orbitals. Because the energies of N 2p and O 2p are not the same, the orbitals are drawn at different heights. From combinations of the 2p orbitals we form σ and π bonding orbitals. These are lower in energy than either of the 2p levels. We can also form antibonding combinations, σ and π, and these lie higher in energy than either 2p level. As with the homonuclear diatomics, the σ bonding orbital is lower than the π bonding orbital and the σ antibonding orbital higher in energy than the π antibonding orbital.

Apart from the slightly different label, there is another difference between the orbitals of heteronuclear molecules, such as NO, and those of homonuclear molecules, such as N_2. For N_2, combinations such as $(2p_{zA} + 2p_{zB})$ had equal contributions from orbitals on atoms A and B. For NO the contributions of atomic orbitals from N and O are not

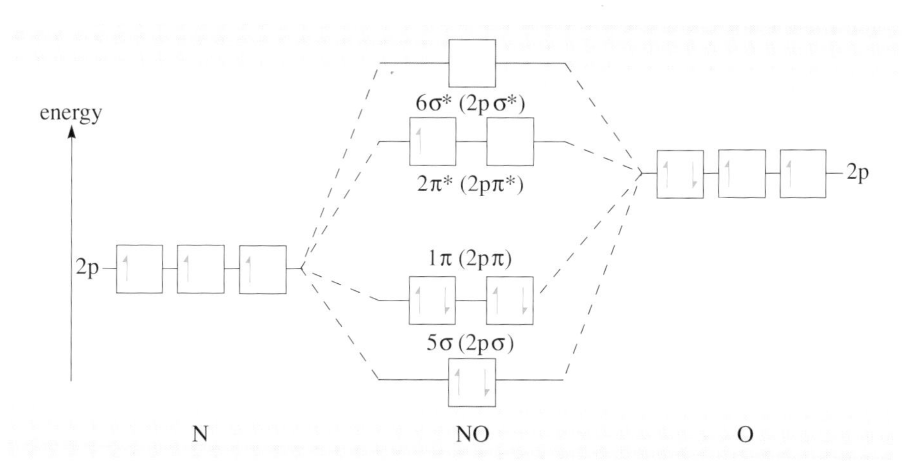

Figure 62 The partial orbital energy-level diagram of NO. The molecular orbitals are labelled in the two ways that we introduced for homonuclear diatomic molecules. Note that we reach higher numbers when labelling the σ and π orbitals thanwe did for N_2 or O_2. This is because we can no longer distinguish g and u, and so have to label all σ and all π orbitals consecutively.

equal. In general the contribution of an atomic orbital will depend on how close in energy it is to the molecular orbital. The σ and π bonding orbitals in NO are closer in energy to the lower N 2p than to the O 2p and these orbitals contain a greater contribution from N 2p. We can represent this as $(2p_{zN} - f2p_{zO})$, where f is less than 1. The antibonding orbitals have a greater contribution from O 2p than from N 2p, so that the antibonding $2p\sigma^*$ can be represented as $(f'2p_{zN} + 2p_{zO})$.

On Figure 62 we have labelled the molecular orbitals in the two ways that we introduced for homonuclear diatomic molecules. Note that we reach higher numbers when labelling the σ and π orbitals (for example 2π and 5σ) than we did for N_2 or O_2. This is because we can no longer distinguish g and u, and so have to label all σ (or π) orbitals consecutively.

■ From Figure 62, what is the bond order of NO?

□ It is $2\frac{1}{2}$. The 1s and 2s orbitals are non-bonding and so do not contribute to the bond order. There are six electrons in the bonding orbitals formed by the 2p orbitals and one in the antibonding orbital $2\pi^*$, giving a bond order of $(6 - 1)/2 = 2\frac{1}{2}$.

■ How would you expect the bond length and dissociation energy of NO to compare with those of N_2 and O_2?

□ N_2, NO and O_2 all use similar orbitals for bonding. Their bond orders are 3, $2\frac{1}{2}$ and 2, respectively, and so we would expect the bond length and dissociation energy of NO to lie between those of N_2 and O_2. The experimental data for the three molecules given in Table 2 support this prediction.

Table 2 A comparison of O_2, NO and N_2

Property	O_2	NO	N_2
bond order	2	$2\frac{1}{2}$	3
number of unpaired electrons	2	1	0
dissociation energy/kJ mol^{-1}	498	632	945
bond length/pm	121	115	109.4

The odd electron in NO is shown in Figure 62 as occupying an antibonding orbital, $2\pi^*$. This suggests that removal of this electron might be fairly easy.

■ What bond order would NO^+ have?

□ Three, the same as N_2. We would expect the NO^+ ion to have a very large dissociation energy like N_2, and larger than that of NO.

Is there any evidence for the existence of NO^+? Yes. The nitrosyl halides will react with other halides to form crystalline salts of NO^+; for example

$$NOCl(g) + SbCl_5(l) = NO^+ SbCl_6^- (s)$$

$$NOF(g) + BF_3(g) = NO^+ BF_4^- (s)$$

These salts can be isolated and their structures determined by X-ray diffraction. They are strong oxidizing agents and react with water, so they cannot be studied in aqueous solution, although NO^+ is found in concentrated sulphuric acid solution. Under certain circumstances, NO will also bind to metal ions, and in many of these compounds will bind as NO^+.

Molecular orbital theory then explains why NO is stable with respect to N and O atoms, and predicts that NO^+ will be formed fairly easily. We have not explained why nitric oxide gas consists of NO molecules rather than N_2O_2 molecules, or why NO is unstable with respect to oxidation by oxygen. To do this, we would have to consider orbital energy-level diagrams for N_2O_2 or NO_2 and O_2 as well.

The other molecule that we shall study in this Section is CO. As we saw for C_2, the carbon 2s and 2p orbitals are fairly close in energy. For CO, then, we would expect that when we combine the 2p atomic orbitals on C and O to form molecular orbitals, we shall have to consider the effect of contributions from the carbon 2s.

■ When we combine the 2p atomic orbitals from C and O, we obtain six molecular orbitals (5σ, 1π (two), 2π (two) and 6σ), as we did by combining 2p orbitals in NO. What effect will mixing in some carbon 2s orbital have?

□ In C_2, the effect of the 2s orbital on the molecular orbitals made from the 2p atomic orbitals was to raise the energy of the $3\sigma_g$ ($2p\sigma_g$) orbital so that it was above the $1\pi_u$ ($2p\pi_u$) orbital. Similarly, the effect of the 2s orbital in CO is to raise the energy of the 5σ ($2p\sigma$) orbital above that of the 1π ($2p\pi$).

In other words, the 5σ orbital is no longer simply a combination of 2p on carbon and 2p on oxygen, but contains some carbon 2s orbital as well. Similarly, the $4\sigma*$ orbital will contain some carbon and oxygen 2p. The orbital energy-level diagram for CO is shown in Figure 63.

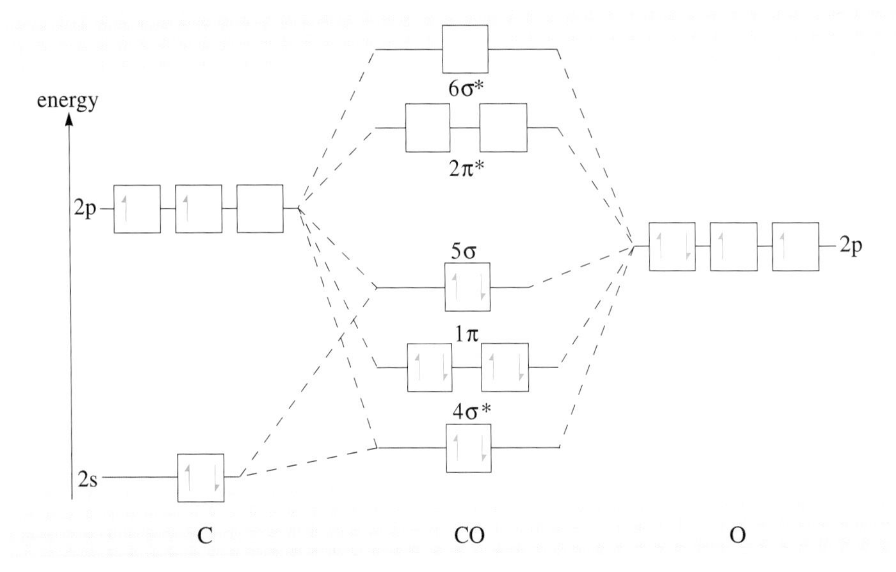

Figure 63 Orbital energy-level diagram for CO.

In molecules in which both the 2s and 2p orbitals on one or both of the atoms have to be considered, such as CO, molecular orbitals cannot be labelled simply bonding and antibonding as they were for N_2 and NO. For example, 5σ in CO consists of the $2p_z$ on oxygen combined with the $2p_z$ on carbon in a bonding manner, but the 2s on carbon is mixed in such a way as to overlap with 2p on oxygen in an antibonding manner. For CO the net result is a bonding orbital. This is partly because the 2p orbitals are closer in energy, but also because the 2p electron is more concentrated along the z axis so that $2p_z$ on oxygen will produce a greater overlap with $2p_z$ on carbon than with 2s on carbon. The simple concept of bond order breaks down for examples like this.

Molecular orbital theory predicts that CO has a very similar electronic structure to N_2, with which it is **isoelectronic**; that is, CO and N_2 have the same number of electrons. CO and N_2 have different bond lengths and molar bond enthalpies (113 pm and 110 pm and 1 076 kJ mol^{-1} and 945 kJ mol^{-1}, respectively) because the orbitals are not exactly the same in the two molecules.

The electrons in the molecular orbitals of CO, unlike those in N_2, are not evenly distributed between the atoms. Four electrons are transferred from O 2p orbitals to molecular orbitals that are mixtures of O 2p and C 2p and C 2s orbitals. Two electrons are transferred from C 2p orbitals to these mixed molecular orbitals.

If the molecular orbitals were equal mixtures of C 2p and O 2p, then the net result would be a transfer of electron density from oxygen to carbon because four electrons came from oxygen and two from carbon, but in the molecule all six would be equally shared between oxygen and carbon. However, because the O 2p orbitals are lower in energy than the C 2p, the bonding molecular orbitals contain more O 2p than C 2p.

Figure 64 compares similar molecular orbitals in CO and N_2.

We find experimentally that there is almost no transfer of electron density from C to O or from O to C. Thus, the transfer of electrons from O 2p orbitals to molecular orbitals that are mixtures of C 2p, C 2s and O 2p is almost completely compensated by the molecular orbital having more of O 2p than C orbitals in it.

■ Assume that the 1π and 5σ orbitals of CO are two-thirds O 2p and one-third C 2p. What is the net transfer of electrons from C to O or from O to C on forming CO from the separate atoms?

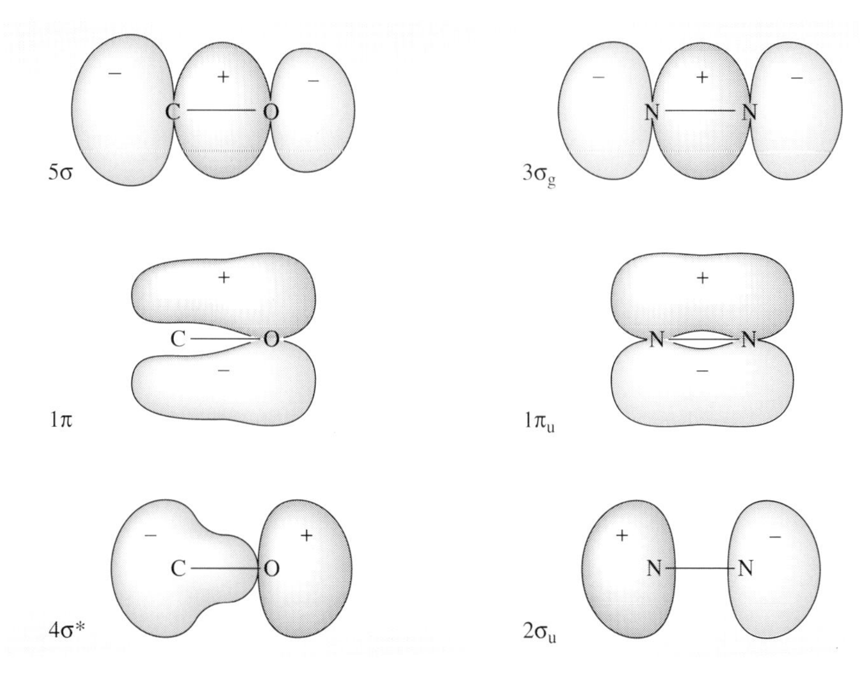

Figure 64 Molecular orbitals for N_2 and CO.

□ Four electrons are transferred from O 2p orbitals to orbitals that are two-thirds O 2p; that is 4 × O 2p becomes $4 \times \frac{2}{3}$ O 2p + $4 \times \frac{1}{3}$ C 2p. Two C 2p electrons are also transferred, to give $2 \times \frac{2}{3}$ O 2p + $2 \times \frac{1}{3}$ C 2p. Thus we began with 4 O 2p and 2 C 2p and ended with $\frac{8}{3}$ O 2p + $\frac{4}{3}$ O 2p = 4 O 2p and $\frac{4}{3}$C 2p + $\frac{2}{3}$ C 2p = 2 C 2p. Our assumptions of the proportions of O 2p and C 2p are simplified, but this illustrates the possibility of forming a stable heteronuclear molecule with no net transfer of electron density from one atom to another.

Carbon monoxide is found as a gas containing the diatomic molecule CO. It is highly poisonous, forming a complex with haemoglobin in the blood and thus preventing oxygen being transported to where it is needed in the body. In this complex, CO is bound to a metal atom (iron) using the 5σ and $2\pi^*$ orbitals. Dinitrogen, however, despite its similar molecular orbital description, does not bind to iron in haemoglobin. The $1\pi_g$ orbital in N_2 is of a less suitable energy for binding to the iron atom than the $2\pi^*$ orbital in CO. Dinitrogen does, however, bind to metal ions in complexes, and attachment of N_2 to a metal ion is believed to occur in enzymes that are involved in making atmospheric nitrogen available to plants.

7.5 SUMMARY OF SECTION 7.4

1 Molecular orbitals for heteronuclear diatomic molecules can be made by combining atomic orbitals of the same symmetry and of similar energy from the two atoms.

2 Combination of two atomic orbitals of the same symmetry produces a bonding orbital of lower energy than either atomic orbital, and an antibonding orbital of higher energy. Further atomic orbitals of the same symmetry can be mixed in by combining them with the resulting orbitals.

3 Because the atomic orbitals combined are not of the same energy, the electron in the molecular orbital is not equally shared between the two atoms.

4 Molecular orbitals in heteronuclear diatomic molecules cannot be labelled with the subscripts g and u because the molecule does not possess a centre of symmetry.

SAQ 32 Draw an orbital energy-level diagram for OF, showing only those orbitals made from the 2p atomic orbitals on both atoms. What is the bond order of OF? Why is this molecule not observed under normal conditions? (The relevant ionization energies are: O 2p, 21.8×10^{-19} J; and F 2p, 27.9×10^{-19} J.)

SAQ 33 Draw an orbital energy-level diagram for CN. Start by combining 2p orbitals on C and N and then consider whether you need to include other orbitals. (The relevant ionization energies are: C 2s, 26.58×10^{-19} J; C 2p, 18.04×10^{-19} J; N 2s, 32.57×10^{-19} J; N 2p, 23.28×10^{-19} J.)

CN is known only from spectroscopic studies, but the cyanide ion CN^- is very well known. With which molecule discussed in this Section is it isoelectronic? Does molecular orbital theory suggest that CN^- would bind to iron?

7.6 SIGMA AND PI BONDING AT LARGE

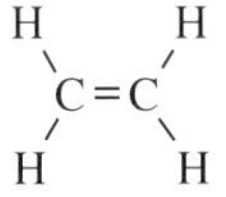

On strict symmetry grounds, the labels σ and π can be used only for molecules belonging to $\mathbf{C}_{\infty v}$ or $\mathbf{D}_{\infty h}$, that is linear molecules. However, it has become customary for chemists to refer to any orbital that binds atoms together by increasing the electron density between nuclei as a σ bonding orbital and any orbital which binds above and below the nuclei as a π orbital. Thus for example, in the molecule ethene (see margin), we talk about the single C—H bonds as being sigma bonds and the C=C double bond as being composed of a σ and a π bond. There is only one π bonding orbital in this case however. This is composed of the C 2p orbitals that point out of the plane of the molecule. The 2p orbitals that would have made the other π_u orbital in C_2 are used to form the C—H bonds. Another example is BF_3, which is a planar molecule. The 2p orbitals in the plane of the molecule form σ orbitals and those out of the plane form π orbitals. The π bonding orbital is shown in Figure 65.

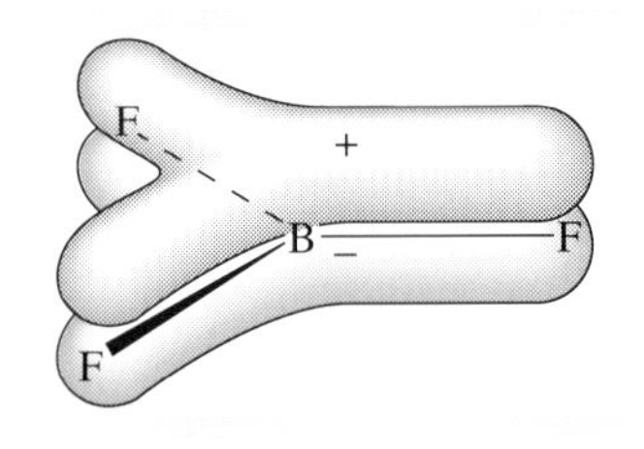

Figure 65 The π bonding orbital in BF_3.

These sigma and pi bonding orbitals have antibonding counterparts. Often the antibonding orbitals are not occupied and so are ignored by chemists interested in the molecule in its lowest energy state. If we want to distinguish between bonding and antibonding orbitals, we denote the antibonding orbital by an asterisk.

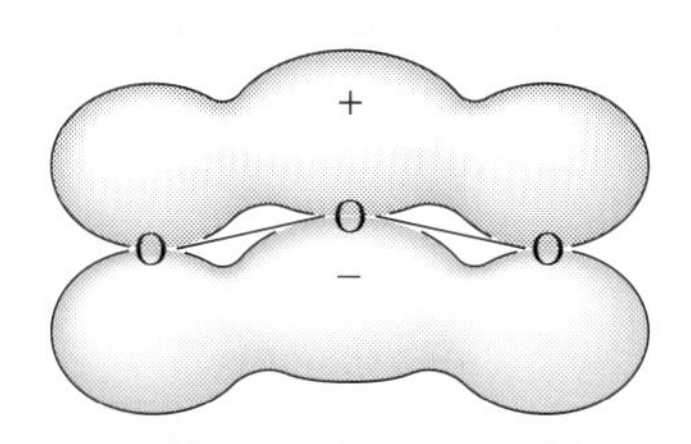

Figure 66 Bonding orbital in O_3.

SAQ 34 Figure 66 shows a bonding orbital in ozone, O_3. Would this be referred to as a σ orbital or a π orbital?

8 MOLECULAR ORBITALS FOR POLYATOMIC MOLECULES

In diatomic molecules, it was fairly easy to decide which atomic orbitals to combine by inspection. With larger molecules, such as ethene or SF_6, it is not always straight forward. The job of deciding which orbitals to combine is simplified by the use of symmetry arguments. There is a general rule that *only orbitals that behave in the same way under the operations of the point group of the molecule can combine to form molecular orbitals*. You have met an instance of this already, where the 2s orbital on carbon in CO behaved like a σ orbital and so could combine with 2p σ but not 2p π.

We shall illustrate how to form molecular orbitals for polyatomic molecules using a very simple example, the water molecule (H_2O), and then see whether the results tie in with our notions of electron-pair bonding and with experiment. But first we need to clarify what is meant by the behaviour of an orbital under the operations of the point group.

8.1 REPRESENTATIONS OF POINT GROUPS

In order to belong to a particular group, a molecule must appear unchanged by all the operations of that group. However, orbitals of the molecule are not so restricted. In homonuclear diatomic molecules, such as N_2 and O_2, for example, the molecule itself has a centre of symmetry but the σ_u and π_u orbitals were changed into minus themselves by inversion through this centre.

In H_2O the obvious orbitals to combine on energy grounds are the 2p on oxygen (ionization energy 21.81×10^{-19} J) and the 1s on hydrogen (21.74×10^{-19} J). Now, if the hydrogen 1s orbitals are to combine with the oxygen 2p then they must behave the same way under the symmetry operations of the point group of H_2O.

■ To which symmetry point group does H_2O belong?

□ C_{2v} .

■ What are the symmetry operations of C_{2v}?

□ $\hat{C}_2 + 2\ \hat{\sigma}_v$: rotation about a two-fold axis and reflection through the two vertical planes of symmetry.

Figure 67 shows the symmetry elements of H_2O.

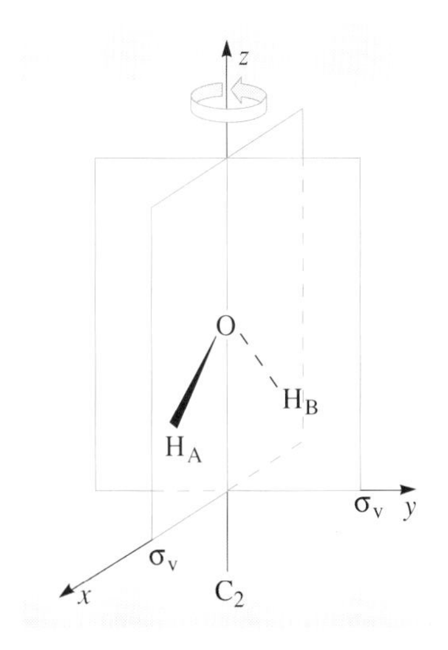

Figure 67 Symmetry elements and molecular axes of H_2O.

How do the 2p orbitals on the oxygen atoms behave? First we define x, y and z axes for the molecule. It has become standard to define the principal axis as the z axis (in this case the C_2 axis) and the xz plane as a vertical plane of symmetry. The

Figure 68 (a) A $2p_z$ orbital on oxygen in H_2O. (b) A $2p_x$ orbital on oxygen in H_2O.

H_2O molecule has two planes of symmetry, and we shall define the plane of the molecule as the *xz* plane, as in Figure 67. Let us start with the $2p_z$ orbital (Figure 68a). Rotation about the *z* axis leaves $2p_z$ unaltered. The result of this rotation is thus still $2p_z$. Reflection through the *yz* plane (the plane of symmetry at right-angles to the molecule) also leaves the $2p_z$ orbital unaffected. Likewise the orbital is unaffected by reflection in the plane of the molecule (the *xz* plane). So all the symmetry operations leave $2p_z$ unaffected. Let's try the $2p_x$ orbital. A $2p_x$ orbital on oxygen in H_2O is shown in Figure 68b.

Unlike the $2p_z$ orbital, the $2p_x$ orbital changes sign when it is rotated around the C_2 axis. Thus we can say that the result of performing a rotation through half a revolution about this axis is minus one times $2p_x$, or $-2p_x$. Reflection through the *yz* plane also changes $2p_x$ to $-2p_x$, but reflection in the plane of the molecule leaves $2p_x$ unchanged.

Finally let's see what happens to the $2p_y$ orbital. This is the orbital with one lobe above the plane of the molecule and a lobe of opposite sign below the plane of the molecule. When we rotate $2p_y$ about the C_2 axis, the two lobes change positions and so we obtain minus one times $2p_y$, or $-2p_y$. Reflection through the *yz* plane, however, leaves the $2p_y$ orbital unchanged. This plane bisects both lobes of $2p_y$. Being above and below the molecular plane, $2p_y$ is turned into minus itself by reflection through the molecular plane ($\hat{\sigma}_v(xz)$).

Now we can summarize the different behaviour of $2p_x$, $2p_y$ and $2p_z$, in tabular form (Table 3). Along the top we write the symmetry operations of the point group. Down the left-hand side we write $2p_x$, $2p_y$ and $2p_z$. Then we fill in what these orbitals change to when acted on by the various operations.

Table 3

	$\hat{C}_2$	$\hat{\sigma}_v(yz)$	$\hat{\sigma}_v(xz)$
$2p_z$	$2p_z$	$2p_z$	$2p_z$
$2p_x$	$-2p_x$	$-2p_x$	$2p_x$
$2p_y$	$-2p_y$	$2p_y$	$-2p_y$

This table can be made more general. We have considered only 2p orbitals, but 3p and 4p orbitals would behave in the same way, and any s orbital on oxygen would behave in the same way as $2p_z$. (Try this for yourself. An s orbital is spherical. Draw a circle round the O atom in H_2O and work out what happens to it when it is acted on by the three operations of $\mathbf{C}_{2v}$.)

We can leave out $2p_x$, $2p_y$ and $2p_z$ in our table and write in just the effect of the operation. $2p_z$ is unchanged by rotation about the C_2 axis and so the result of acting on $2p_z$ with $\hat{C}_2$ is $2p_z$, that is plus one times $2p_z$, and so we put a 1 under $\hat{C}_2$ for $2p_z$. Similarly $2p_x$ is changed by $\hat{C}_2$ into $-2p_x$, that is minus one times $2p_x$, and so we put a -1 under $\hat{C}_2$ for $2p_x$. The resulting table is given in Table 4.

Table 4

$\hat{C}_2$	$\hat{\sigma}_v(yz)$	$\hat{\sigma}_v(xz)$
1	1	1
−1	−1	1
−1	1	−1

The three rows of numbers in this table represent three patterns of behaviour. The first one is that illustrated by p_z and s orbitals on oxygen for example, and the second one that obeyed by all p_x orbitals on oxygen, and the third is that exhibited by all the p_y orbitals.

Now, it turns out that all the atomic and molecular orbitals in H_2O can be represented by four such patterns. These four basic patterns are called **irreducible representations**. All molecular orbitals of molecules belonging to the group $\mathbf{C}_{2v}$ will behave like one of the four irreducible representations of that group.

■ The fourth representation is illustrated by a d_{xy} atomic orbital on sulphur in H_2S, as shown in Figure 69. Write down the fourth pattern or irreducible representation by considering the behaviour of this orbital.

□ Under C_2, the orbital swings around like a propeller but ends up with positive lobes and negative lobes in the same places as at the start. The result is thus $1 \times d_{xy}$. Reflection through the $\sigma_v(yz)$ plane swaps the top negative lobe with the top positive lobe and the bottom negative lobe with the bottom positive lobe giving $-1 \times d_{xy}$. $\hat{\sigma}_v(xz)$ also swaps positive and negative lobes; in this case a top lobe with a bottom lobe. The irreducible representation is therefore:

$\hat{C}_2$	$\hat{\sigma}_v(yz)$	$\hat{\sigma}_v(xz)$
1	−1	−1

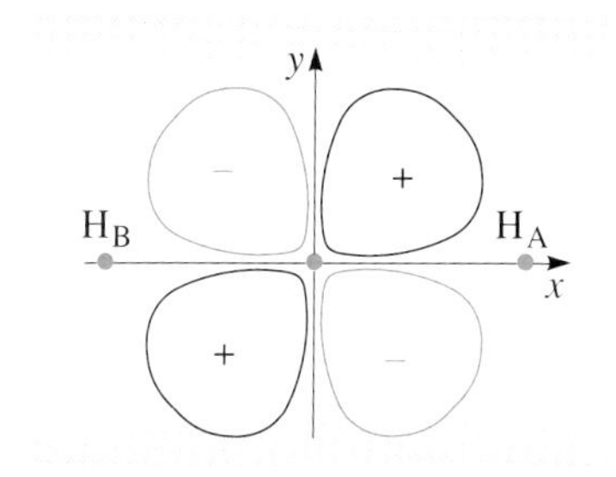

Figure 69 A d_{xy} orbital on sulphur in H_2S, viewed vertically down the z axis.

8.1.1 BUSY DOING NOTHING

Tables of behaviour patterns like the one we have just considered have been worked out and published for large numbers of symmetry point groups. Before you can use these tables however, you need to meet another symmetry operation. This operation, called the **identity operation**, is the act of doing nothing. For our purposes we can regard it as a device for keeping track of how many orbitals we are considering. Thus, in the examples above, in each case we took only one 2p orbital on oxygen. These irreducible representations would have a one under the identity operation. If we were considering two orbitals, for example the two hydrogen 1s orbitals in H_2O, then we would have to put 2 under the identity operation.

The reason the identity operation is usually included in such tables has to do with the properties of groups. In order for a set of operations to form a group, the result of performing two of the operations (or one operation twice) must be the same as that produced by performing one of the group operations. Now, reflecting through a plane of symmetry twice is not the same as rotation through half a revolution, or the same as reflecting through a plane of symmetry once: it is the same as doing nothing, and so we have to include the operation of doing nothing.

All symmetry point groups contain the identity operation and it is given the symbol $\hat{I}$ (or sometimes $\hat{E}$).

The complete table of behaviour patterns for $\mathbf{C}_{2v}$ is that given in Table 5. A table such as this is known as a **character table**.

Table 5

	$\hat{I}$	$\hat{C}_2$	$\hat{\sigma}_v(xz)$	$\hat{\sigma}_v(yz)$
a_1	1	1	1	1
a_2	1	1	−1	−1
b_1	1	−1	1	−1
b_2	1	−1	−1	1

Here we have labelled the four irreducible representations. Irreducible representations with 1 under the identity operation are usually labelled a or b. The exception is for the groups $\mathbf{C}_{\infty v}$ and $\mathbf{D}_{\infty h}$, where such representations are labelled σ. If rotation about the principal axis by $1/n$ of a rotation produces no change, that is there is a 1 under $\hat{C}_n$, then the representation is a. Representations with a −1 under $\hat{C}_n$ are labelled b. In groups with centres of symmetry, if no change is produced by inversion through the centre of symmetry, then the representation is labelled with a subscript g, for example a_g or b_{1g}. Representations with a −1 under $\hat{i}$ are labelled with a subscript u. Subscripts 1, 2, 3, etc. and/or superscripts ′, ″, etc. are used to distinguish behaviour under reflection through planes of symmetry. Thus a_1 and b_1 above have 1 under $\hat{\sigma}_v(xz)$ whereas a_2 and b_2 have −1 under $\hat{\sigma}_v(xz)$. There is no need to remember the rules for labelling irreducible representations; you will always be given character tables that are already labelled.

Now how are the irreducible representations going to help us to form molecular orbitals for water? So far, we have looked only at atomic orbitals on the oxygen atom. We need to combine these with atomic orbitals on the hydrogen atoms, and so we next turn to how we deal with these.

8.1.2 REDUCIBLE REPRESENTATIONS

There are two hydrogen atoms in H_2O and they can be interchanged by operations of the group $\mathbf{C}_{2v}$. We therefore have to consider the two hydrogens together.

Hydrogen has only one electron in a 1s orbital. The energy of this orbital is very close to that of the 2p orbital on oxygen and so we shall look at the hydrogen 1s orbitals. Let's see how two 1s orbitals, one on each hydrogen, behave under the operations of $\mathbf{C}_{2v}$.

There are two 1s orbitals and so under the identity operation we put 2. Under rotation, the two orbitals swap places. They are neither themselves nor minus themselves, but each looks like the other orbital. In this case we put 0 under the operation $\hat{C}_2$. (If you consider the 1s orbital on one atom, which we shall label 1s(A), then the effect of the operation is to move both atom and orbital so that it now looks like a 1s orbital in the other position 1s(B). Thus 1s(A) has been changed to $0 \times$ 1s(A) + $1 \times$ 1s(B). For the character table we only need the amount of 1s(A), and that is 0. Similarly for 1s(B), $\hat{C}_2$ changes it to $1 \times$ 1s(A) + $0 \times$ 1s(B) and we need only the amount of 1s(B).)

The effect of reflection through the yz plane is similar, giving another 0. Finally, reflection through the xz plane leaves both 1s orbitals in the same place and so we put a 2 under $\hat{\sigma}_v(xz)$. The representation of the two hydrogen 1s orbitals is thus as shown in the margin.

$\hat{I}$	$\hat{C}_2$	$\hat{\sigma}_v(xz)$	$\hat{\sigma}_v(yz)$
2	0	2	0

Now this is obviously not one of the patterns in the character table. It is, however, the sum of two of them. To add two irreducible representations together, we add the numbers under each column separately; adding a_1 and a_2 for example gives

	$\hat{I}$	$\hat{C}_2$	$\hat{\sigma}_v(xz)$	$\hat{\sigma}_v(yz)$
a_1	1	1	1	1
a_2	1	1	−1	−1
$a_1 + a_2$	2	2	0	0

■ Which two irreducible representations do you need to add to give the representation of the two hydrogen 1s orbitals?

□ $a_1 + b_1$.

This tells us that the two hydrogen 1s orbitals can combine with a_1 and b_1 orbitals. Before we take the final step and combine atomic orbitals to make molecular orbitals, try these SAQs.

SAQ 35 The molecule OF_2 also belongs to $\mathbf{C}_{2v}$. Consider a $2p_z$ orbital on each fluorine in this molecule, and determine the representation of the two $2p_z$ orbitals by considering the effect of the operations of $\mathbf{C}_{2v}$ on them. Express your result as a sum of irreducible representations.

(Hint: If the atoms swap places then put a 0 under that operation.)

SAQ 36 The molecule SF_4 belongs to $\mathbf{C}_{2v}$. Determine the representation of two 2s orbitals, one on each equatorial fluorine, in this molecule. Which two irreducible representations add together to form this representation? Why do we consider the two equatorial fluorines separately from the two axial fluorines?

SAQ 37 The molecule ethene, C_2H_4, belongs to $\mathbf{D}_{2h}$, whose character table is given in the *Data Book*. Take the z axis along the C=C bond, and the x axis in the plane of the molecule. Check for yourself that all four hydrogen atoms can be interchanged by operations of the group. Determine the representation of four 1s orbitals, one on each hydrogen atom. Give the irreducible representations that you need to add to make this representation.

8.2 MOLECULAR ORBITALS FOR WATER

Now that we have determined the irreducible representations of the atomic orbitals, we can use the fact that only atomic orbitals belonging to the same irreducible representation can be combined to form molecular orbitals.

The two 1s orbitals on hydrogen in H_2O belong to a representation that is the sum of a_1 and b_1. These orbitals, then, can combine only with atomic orbitals on oxygen that belong to a_1 or b_1.

■ To which irreducible representations do the 2p orbitals on oxygen belong?

□ $2p_z$ had 1 under all the symmetry operations and so is a_1; $2p_x$ had −1 under $\hat{C}_2$ and $\hat{\sigma}_v(yz)$ and so belongs to b_1; and $2p_y$ had −1 under $\hat{C}_2$ and $\hat{\sigma}_v(xz)$ and belongs to b_2.

Thus the 1s orbitals can combine with $2p_z$ and $2p_x$ but not with $2p_y$.

$2p_z$ combines with the 1s orbitals as shown in Figure 70, and forms a bonding molecular orbital and an antibonding molecular orbital. Both are labelled a_1. The a_1 bonding orbital is a σ bonding orbital, and it bonds both hydrogens to the oxygen.

Figure 70 Bonding and antibonding combinations of $2p_z$ on oxygen with 1s on hydrogen in H_2O.

The combinations of $2p_x$ with the 1s orbital are shown in Figure 71. Again there are both bonding and antibonding orbitals.

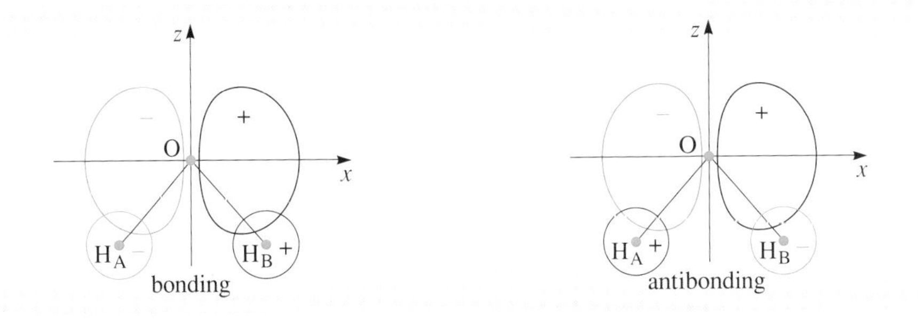

Figure 71 Bonding and antibonding orbitals formed from oxygen $2p_x$ and hydrogen 1s in H_2O.

■ What will be the labels for these molecular orbitals?

□ They will both be labelled b_1 because they are formed from atomic orbitals belonging to b_1.

The b_1 bonding orbital is also a sigma bonding orbital joining both hydrogens to the oxygen, but in this case there are two parts of opposite sign.

The $2p_y$ orbital on oxygen remains uncombined; it becomes a non-bonding π orbital labelled b_2.

Altogether we have formed five molecular orbitals, a_1 bonding, a_1 antibonding, b_1 bonding, b_1 antibonding and b_2 non-bonding, from five atomic orbitals (three 2p orbitals on oxygen and one 1s on each hydrogen). It can be useful to remember the general rule that *n atomic orbitals combine to make n molecular orbitals.*

Here is a summary of the steps that we took to obtain the molecular orbitals of H_2O. We shall see how these can be applied to other molecules in the next Section.

1 Decide the shape of the molecule. In the case of H_2O we assumed the experimentally determined shape. If no experimental data are available, we can use VSEPR to provide a starting point.

2 Determine the symmetry point group of the molecule.

3 Decide which atomic orbitals are close enough in energy to combine. In general the atomic orbitals we have to consider are the valence orbitals.

4 Work out the representation of the atomic orbitals chosen on each atom, or group of equivalent atoms, under the operations of the symmetry point group. For H_2O we considered atomic orbitals on oxygen and atomic orbitals on the two equivalent hydrogen atoms.

5 Combine atomic orbitals that belong to the same irreducible representation to form molecular orbitals.

Finally, having obtained the molecular orbitals, we need to represent them on an orbital energy-level diagram. On one side we show the 2p energy level of oxygen, and on the other side we put two boxes representing the two 1s orbitals, one from each hydrogen. In the middle we put the energy levels of H_2O. The bonding orbitals must lie below both the O 2p and the H 1s. As you will see later, it can be shown that the b_1 bonding orbital is lower in energy than the a_1 bonding orbital. So we put b_1 at the bottom, followed by a_1. Next we have the **non-bonding orbital** b_2. This is purely an O 2p orbital and so has the same energy as it did in the oxygen atom. Lastly, since the b_1 bonding orbital was the more bonding, the b_1 antibonding orbital will be the more antibonding and so will be above the a_1 antibonding level. The orbital energy-level diagram is shown in Figure 72.

■ What is the bond order of H_2O from Figure 72?

□ There are four electrons (two pairs) in bonding orbitals, two in a non-bonding orbital, and none in antibonding orbitals. *Non-bonding orbitals do not count towards bond order*, and so the bond order is 2.

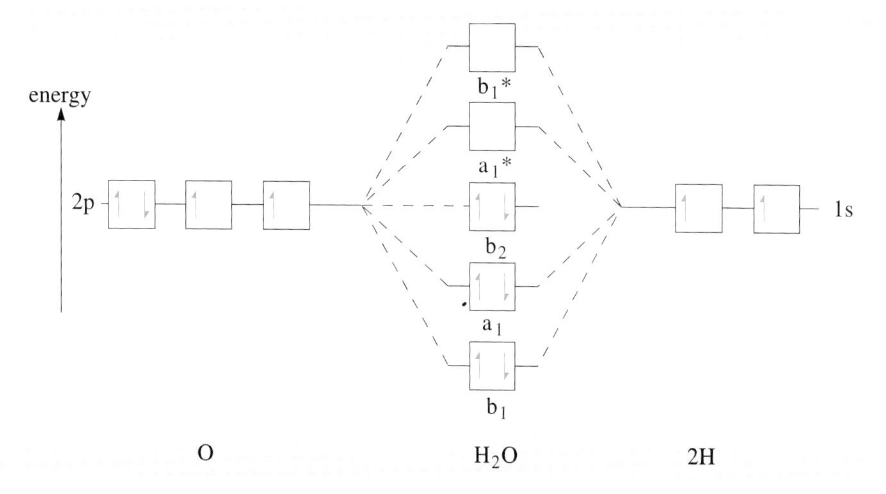

Figure 72 Orbital energy-level diagram for H_2O.

8.3 ORBITALS FOR OTHER MOLECULES

In this Section we look at two other molecules and see how to apply the rules from the previous Section.

8.3.1 METHYLENE, CH_2

The shape of CH_2 has been determined experimentally. It is a bent molecule like H_2O and so belongs to $\mathbf{C}_{2v}$. On energy grounds we considered 2p on oxygen and 1s on hydrogen for H_2O. However, in CH_2, the 2s and 2p on carbon are close in energy so we have to consider both of these and also 1s on hydrogen. We already know to which irreducible representations the 2p on carbon and the 1s on hydrogen belong, because from a symmetry point of view it does not matter whether the central atom is oxygen or carbon.

■ What are the irreducible representations of the three 2p orbitals on carbon and the two 1s orbitals on hydrogen?

□ $2p_z$ belongs to a_1, $2p_x$ to b_1 and $2p_y$ to b_2. The two 1s orbitals on hydrogen belong to $a_1 + b_1$.

We now have to determine the irreducible representation of the 2s orbital on carbon.

■ How does the 2s orbital on carbon behave under the symmetry operations of $\mathbf{C}_{2v}$?

□ It is unchanged by all of them. It therefore has the representation

$\hat{I}$	$\hat{C}_2$	$\hat{\sigma}_v(xz)$	$\hat{\sigma}_v(yz)$
1	1	1	1

and belongs to a_1.

Thus the 2s on carbon can be mixed into only a_1 molecular orbitals. The a_1 bonding orbital will be a combination of $2p_z$ on carbon, the two 1s on the hydrogens *and* the 2s on carbon, and so will differ from that in H_2O. The 2s could also contribute to the a_1 antibonding orbital, but does not contribute to b_1 or b_2 orbitals. The b_1 orbitals will still be combinations of $2p_x$ and 1s on hydrogen, and the $2p_y$ orbital remains a non-bonding b_2 orbital.

■ How will adding in 2s affect the energy of the a_1 bonding orbital?

□ The energy of the a_1 bonding orbital will be raised, as were the energies of the C_2 and CO bonding orbitals.

8.3.2 SULPHUR TETRAFLUORIDE, SF_4

SF_4 has the unusual shape you saw in Section 3. Despite its very different appearance, it too belongs to the group $\mathbf{C}_{2v}$. On energy grounds, we consider 2p orbitals on fluorine and 3p orbitals on sulphur. We now have to decide how these orbitals behave under the operation of $\mathbf{C}_{2v}$. First, we must fix the x, y and z axes. The z axis is along the C_2 axis of symmetry, and we shall place the x axis along the axial S—F bonds. The axes are shown in Figure 73.

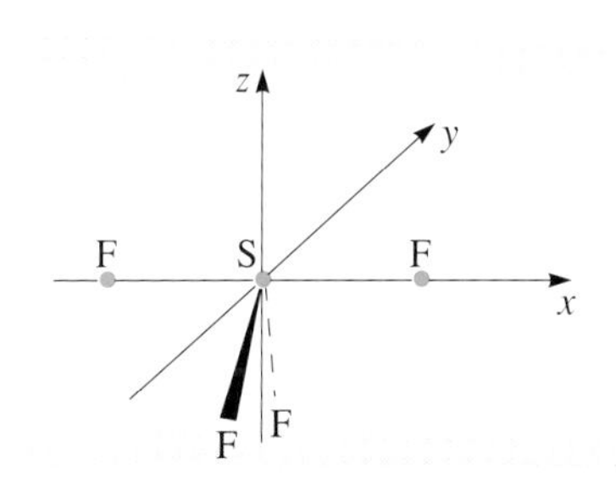

Figure 73 Axes for SF_4.

First, we take the 3p orbitals on sulphur. From a symmetry point of view, 3p orbitals behave like 2p orbitals.

■ What are the behaviour patterns of the 3p orbitals on S in SF_4?

□ The representation is as follows:

	$\hat{I}$	$\hat{C}_2$	$\hat{\sigma}_v(xz)$	$\hat{\sigma}_v(yz)$
$3p_z$	1	1	1	1
$3p_x$	1	−1	1	−1
$3p_y$	1	−1	−1	1

This means that $3p_z$ belongs to a_1, $3p_x$ to b_1 and $3p_y$ to b_2. The p orbitals on the central atom behave the same way as those in H_2O.

The two axial fluorines can be interchanged by symmetry operations of the molecule and so we treat these together. Axial and equatorial fluorines are not interchanged by any operations of $\mathbf{C}_{2v}$ and so we treat the equatorial fluorines separately from the axial fluorines. Let's take the axial fluorines first.

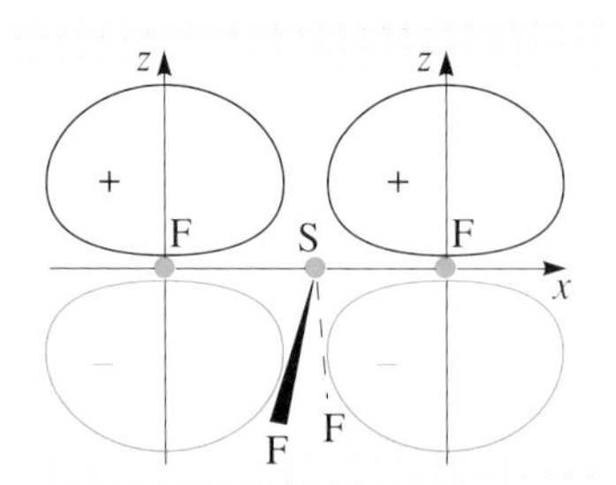

Figure 74 $2p_z$ orbitals on the axial fluorines in SF_4.

The $2p_z$ orbitals on the axial fluorines are as shown in Figure 74. There are two orbitals so we put 2 under $\hat{I}$. Rotation through 180° about C_2 swaps the two fluorines, so under $\hat{C}_2$ there is a 0. Reflection in the yz plane (see Figure 73) also swaps the two fluorines and so we have another 0. Finally, reflection in the xz plane, which is the plane containing sulphur and the two axial fluorines, leaves the $2p_z$ orbitals unaltered and so we put 2 under $\hat{\sigma}_v(xz)$.

There are also two $2p_x$ orbitals as shown in Figure 75. Since the two fluorines are swapped by $\hat{C}_2$ and $\hat{\sigma}_v(yz)$, all orbitals on these atoms get 0 under these operations. So we have 0 under $\hat{C}_2$ and $\hat{\sigma}_v(yz)$ for the $2p_x$ orbitals. Reflection through the xz plane leaves the orbitals unchanged and since there are two of them, we put +2 under $\hat{\sigma}_v(xz)$. Finally two $2p_y$ orbitals on the fluorines will also have 2 under $\hat{I}$, and 0 under $\hat{C}_2$ and $\hat{\sigma}_v(yz)$. However, $\hat{\sigma}_v(xz)$ swaps the positive and negative lobes and so we have to put −2 under this. The behaviour of all the 2p orbitals is summarized in Table 6.

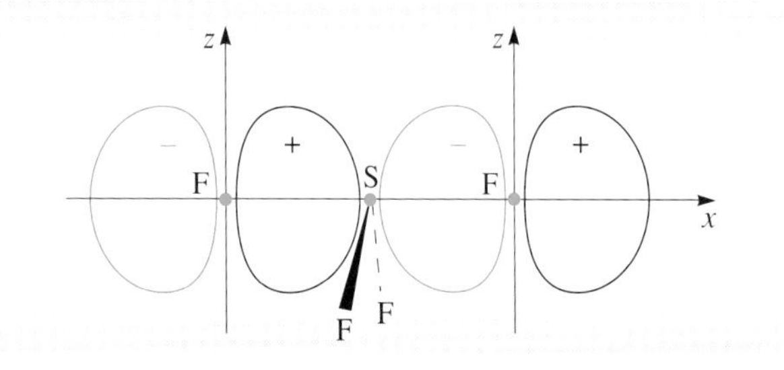

Figure 75 The $2p_x$ orbitals on the axial fluorines in SF_4.

Table 6

	$\hat{I}$	$\hat{C}_2$	$\hat{\sigma}_v(xz)$	$\hat{\sigma}_v(yz)$	
$2p_z$	2	0	2	0	$a_1 + b_1$
$2p_x$	2	0	2	0	$a_1 + b_1$
$2p_y$	2	0	−2	0	$a_2 + b_2$

A similar analysis for the 2p orbitals on the equatorial fluorines shows that the $2p_z$ orbitals belong to $a_1 + b_2$ as do the $2p_y$. The $2p_x$ orbitals belong to $a_2 + b_1$.

Because there are three p orbitals on each atom in SF_4 there are fifteen atomic orbitals to combine to make molecular orbitals. From fifteen atomic orbitals we have to make fifteen molecular orbitals. The use of symmetry, as above, allows us to reduce this problem somewhat. It tells us that the fifteen molecular orbitals comprise five labelled a_1, four labelled b_2, four labelled b_1 and two labelled a_2. Furthermore, we know which five atomic orbitals contribute to the a_1 molecular orbitals, and similarly for a_2, b_1 and

b_2, and we can deal with each set separately. This example thus illustrates how we can reduce the problem of deciding which atomic orbitals to combine in larger molecules (providing the molecules are sufficiently symmetric).

We shall not expect you to analyse a molecule as complex as SF_4, but to see if you have grasped the principles of the method, try the following SAQs.

SAQ 38 Obtain the reducible representations of the 2p orbitals on the equatorial fluorines in SF_4, and show that the $2p_z$ and $2p_y$ belong to $a_1 + b_2$.

SAQ 39 The ozone molecule, O_3, is a bent molecule. Use the method described in this Section to determine which atomic orbitals will combine to form molecular orbitals in ozone.

SAQ 40 Determine the irreducible representations of the molecular orbitals in H_2Se that are formed from hydrogen 1s and selenium 4p orbitals.

SAQ 41 The representations of the 2p orbitals on the two carbons in ethene are:

	$\hat{I}$	$\hat{C}_2(z)$	$\hat{C}_2(y)$	$\hat{C}_2(x)$	$\hat{i}$	$\hat{\sigma}_h(xy)$	$\hat{\sigma}_v(xz)$	$\hat{\sigma}_v(yz)$
$2p_z$	2	2	0	0	0	0	2	2
$2p_x$	2	−2	0	0	0	0	2	−2
$2p_y$	2	−2	0	0	0	0	−2	2

To which irreducible representations do each of these correspond? (Determine the representation of the 1s orbitals on the hydrogen atoms, and decide which carbon 2p orbitals contribute to molecular orbitals that bond carbon to hydrogen. The axes are as given in Figure 76.)

Figure 76 The molecular axes of ethene.

8.4 ENERGY-LEVEL DIAGRAMS FOR POLYATOMIC MOLECULES

8.4.1 WATER

So far, the relative energy levels corresponding to molecular orbitals for polyatomic molecules, such as H_2O, have been discussed only briefly. We can use the bonding, non-bonding or antibonding properties of the orbitals as a starting point. Bonding orbitals lie at lower energy than the atomic orbitals they are made from, and antibonding orbitals higher. Non-bonding orbitals remain at the atomic energy level.

So for H_2O we had a_1 and b_1 bonding orbitals. These will be lower in energy than both the 2p in oxygen and the 1s in hydrogen. The a_1 and b_1 antibonding orbitals will be higher in energy than either atomic level, and the b_2 non-bonding orbital will remain at the energy of the 2p in oxygen.

What remains to be determined is which of a_1 and b_1 is the lower energy bonding orbital. This can be determined only by actually calculating the energies of these orbitals, but we can get an idea by considering the amount of overlap there is between the 2p and the 1s in the two molecular orbitals. The greater the overlap, the more strongly bonding and lower in energy is the orbital.

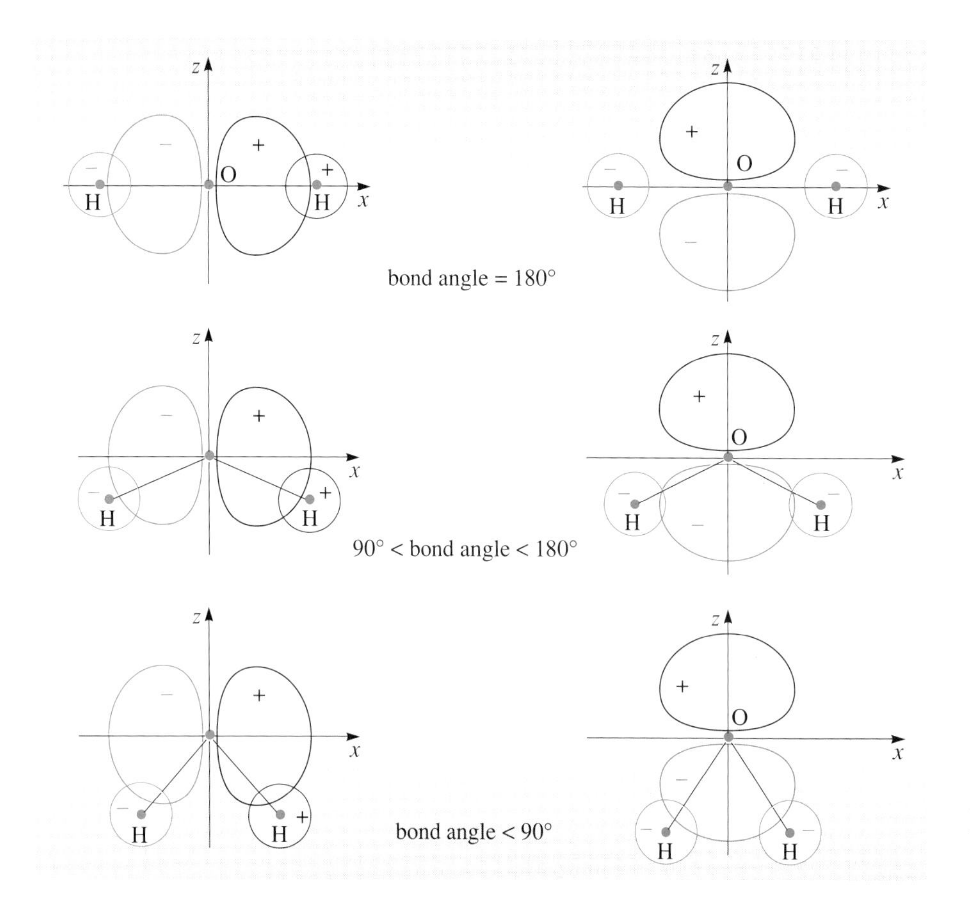

Figure 77 Changing overlap of H 1s with O $2p_x$ (left) and O $2p_z$ (right) as the H_2O molecule is bent.

The amount of overlap depends on the bond angle. If H_2O were linear, there would be a large overlap for $2p_x$ and no net overlap for $2p_z$. As the bond angle decreases, the overlap of the H 1s with $2p_x$ decreases and that with $2p_z$ increases (Figure 77). With a bond angle of 90°, the two overlaps are equal, and for an angle of less than 90° the overlap with $2p_z$ is greater than that with $2p_x$. The bond angle in H_2O is actually 104.5°, so there will be more overlap of the H 1s with the O $2p_x$ than with the O $2p_z$.

■ Which molecular orbital in H_2O will be lower in energy, a_1 or b_1?

□ A large overlap means that there is a high probability of finding the electron between the nuclei. The larger the overlap, therefore, the better the electron in that orbital is at binding the nuclei together and the lower in energy is the orbital. For H_2O, the overlap of O $2p_x$ with H 1s is larger than the overlap of O $2p_z$ with H 1s, and so the b_1 orbital will be lower in energy (more bonding) than the a_1.

Now we can construct an energy-level diagram for H_2O. The bonding orbitals will be lowest in energy and, as we have seen, b_1 will be lower than a_1. The b_2 non-bonding orbital will have the same energy as the oxygen 2p. The a_1 and b_1 antibonding orbitals will be highest in energy. Figure 78 shows the energy-level diagram. In this diagram we have used lines rather than boxes to represent orbitals. This is how energy levels are usually represented, and we shall adopt this style from now on.

Each hydrogen has one electron and the oxygen has eight electrons. The 1s and 2s orbitals on oxygen each contain two electrons and are essentially non-bonding. There are, therefore, six electrons to fit into the molecular orbitals shown in Figure 78. Two each go into the bonding orbitals a_1 and b_1 and, as we saw earlier, these form the O—H bonds in H_2O. The last two go into the non-bonding b_2 orbital. We can think of these two as forming a lone pair.

If you think back to the Lewis structure of H_2O, however, you will remember that in this theory, H_2O had two single O—H bonds and two lone pairs. Are there two lone pairs in molecular orbital theory?

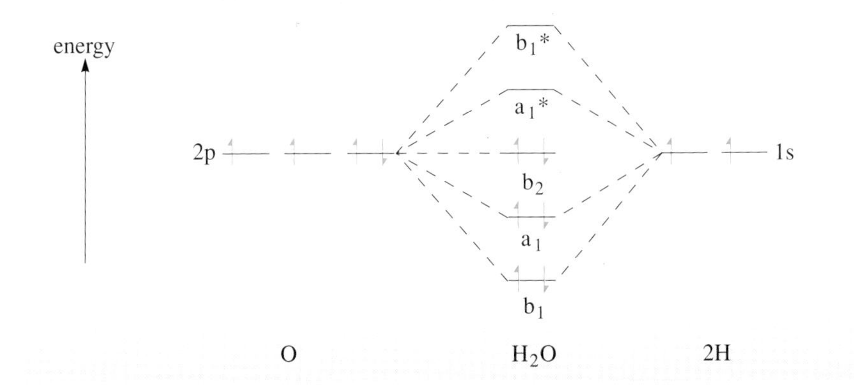

Figure 78 Energy-level diagram for H_2O.

■ Which electrons around oxygen did we use for the Lewis structure?

□ 2s and 2p.

The 2s in molecular orbital theory formed a non-bonding orbital, lower in energy than the orbitals shown in Figure 78. The electrons in this orbital are the second lone pair. The two lone pairs on the molecular orbital theory are in orbitals with very different energies, and so we might look for some experimental results to back up this conclusion.

There is a technique, known as photoelectron spectroscopy, that enables us to measure the energies of electrons in molecules. High energy radiation (usually ultraviolet radiation or X-rays) knocks an electron out of the molecule. The energy of the electron is then measured. If the electron is strongly bound to the molecule, that is it is in a low energy level, then a large amount of energy is needed to knock the electron out and so there will be very little energy left to form the kinetic energy of the electron and the electron will move slowly. An electron in a higher energy level will take less energy to be removed and so will move faster because more of the energy of the radiation is available to it as kinetic energy. So for example, if we shone ultraviolet radiation on H_2O molecules, we would expect the 2s electrons that escape to have a lower velocity than the electrons from the b_2 orbital. The photoelectron spectrum of H_2O shows that there are three electron energy levels close to oxygen 2p as in Figure 78 and one close to the oxygen 2s level. This is as expected from molecular orbital theory but not what one might predict from a simple Lewis structure, which has two equivalent O—H bonds and two equivalent lone pairs.

Energy-level diagrams like Figure 78 can be constructed for almost any molecule, but will become more and more complex as the number of atoms increases. Even the diagram for SF_4, with fifteen molecular orbitals, is rather cluttered and so we shall deal only with very simple molecules as examples. We shall discuss CH_2 and then let you try some examples yourself in the SAQs.

8.4.2 METHYLENE, CH_2

With CH_2 we are still concerned with the a_1 and b_1 bonding, b_2 non-bonding and a_1 and b_1 antibonding orbitals as in H_2O, but we saw earlier that the a_1 had a contribution from the 2s on carbon.

As in C_2 and CO, the mixing of 2s and an orbital involving 2p leads to a lowering of 2s and raising of 2p. In this case, then, the relative energies of a_1 and b_1 depend not only on the bond angle but also on the addition of 2s to a_1 but not to b_1.

For an angle between 90° and 180°, a_1 lies above b_1 and the addition of 2s pushes it closer to b_2, and so the diagram for CH_2 will be similar to that for H_2O but with a_1 higher in energy. This makes the a_1 and b_2 levels so close in energy that the most stable state of the molecule is one with one electron in a_1 and one in b_2 (Figure 79).

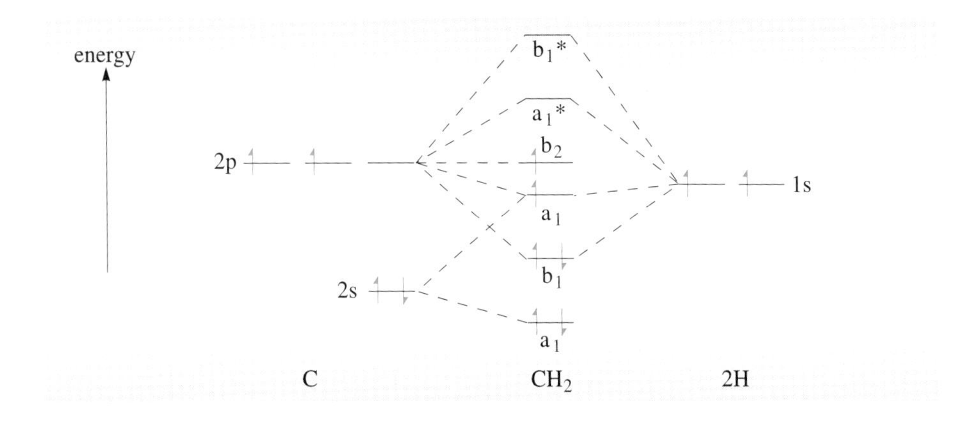

Figure 79 Energy-level diagram for CH_2.

■ Use Figure 79 to predict the magnetic properties of CH_2.

□ CH_2 has two unpaired electrons, one in a_1 and one in b_2. These have parallel spins and so the molecule is paramagnetic.

SAQ 42 Draw an energy-level diagram for H_2S showing 3p orbitals on S, 1s orbitals on H and the molecular orbitals formed by combining these.

It is sometimes suggested that the S—H bonds in H_2S include a contribution from 3d orbitals on sulphur. Consider a $3d_{xz}$ orbital on sulphur (Figure 80) and determine its irreducible representation. How would this orbital affect those shown in your energy-level diagram?

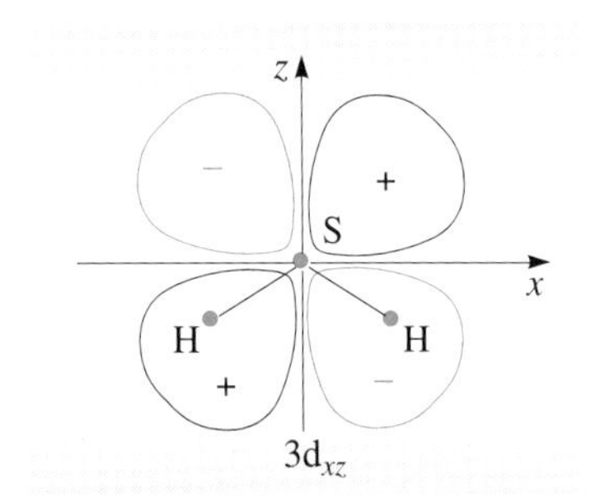

Figure 80 A $3d_{xz}$ orbital on S in H_2S.

SAQ 43 Construct energy-level diagrams for the molecule O_3, ozone, using orbitals formed from oxygen 2p. Treat π orbitals (above and below the plane of the molecule) and σ orbitals (in the molecular plane) separately. Take the molecular plane as the xz plane. (Separation into π and σ makes the diagrams less cluttered).

8.5 LOCALIZED ORBITALS AND THE RELATIONSHIP OF MOLECULAR ORBITALS TO OTHER BONDING DESCRIPTIONS

In the water molecule, the a_1 and b_1 bonding orbitals and the b_2 non-bonding orbitals are full; they each contain two electrons.

As shown above, the a_1 and b_1 orbitals bond all three atoms together. It is possible to reorganize the electron density in these orbitals so that it can be seen to resemble two O—H bonds. We do this by combining the a_1 and b_1 orbitals. The resulting combinations are shown in Figure 81. These orbitals are localized orbitals because an electron in them is localized in one bond, not spread over the whole molecule. It is possible to rearrange the molecular orbitals like this only if either (a) the orbitals have the same energy or (b) both orbitals are equally full. In the case of H_2O it is the latter condition that holds. Such rearrangements can often be used to produce a picture of electron density that ties in with ideas of bonds between atoms, lone pairs, etc. We have seen that we can rearrange the electron density in the bonding orbitals of H_2O into something resembling the O—H σ bonds. We can also combine the electron density in 2s and $2p_y$ to form two lone pairs.

Figure 81 Combinations of a_1 and b_1 orbitals in H_2O.

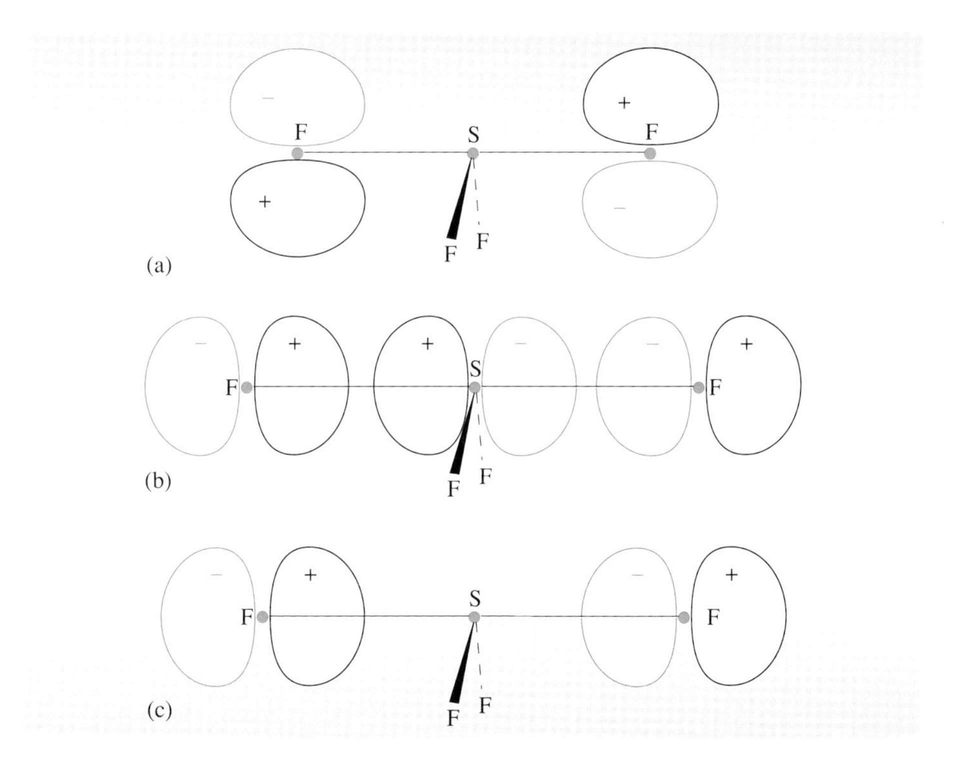

Figure 82 Some orbitals in SF_4: (a) b_1 non-bonding; (b) b_1 bonding; (c) a_1 non-bonding. (The S—F bond lengths have been exaggerated for clarity.)

■ Can we combine a_1 and b_1 bonding orbitals in CH_2?

□ No. They have different energies and are not equally occupied: b_1 contains two electrons but a_1 contains only one.

In CH_2 it is better to stay with the delocalized orbital picture, with all three atoms bound by one electron-pair bond plus an odd electron, than to think of the bond in terms of one C—H bond and one C—H half-bond.

To conclude this brief look at polyatomic molecules, we examine one or two examples of SF_4 bonding orbitals and see how we would describe them in more familiar terms. The orbital shown in Figure 82a is a b_1 non-bonding orbital.

■ Why is it described as non-bonding rather than antibonding?

□ The two p orbitals that contribute are on two different F atoms that are separated by a sulphur atom and so are a long way apart. There is little overlap of the orbitals and they remain essentially atomic.

We can think of electrons in this orbital as contributing to lone pairs on the fluorine atoms.

Figure 82b shows a b_1 bonding orbital. If we combine only p orbitals on sulphur and fluorine, this is the only σ bonding orbital between the sulphur and the axial fluorines. Two electrons in this orbital form a **three-centre bond** covering the sulphur and both fluorines. The a_1 combination of the two $2p_x$ orbitals, Figure 82c, does not form a bonding overlap with any 3p orbital on S and so remains non-bonding, contributing to the lone pairs on fluorine rather than the F—S—F bonds.

In contrast to the axial fluorines, the 2p orbitals on the equatorial fluorines combine with $3p_z$ and $3p_y$ on sulphur to form two σ bonding orbitals, as in H_2O.

■ Which would you expect to be stronger, the axial or the equatorial bonds?

□ The equatorial bonds are stronger. This is borne out by the longer axial S—F bond length.

Now try the following SAQs before you move on to Case Study 2 and discover how molecular orbital theory can be applied to crystalline solids.

SAQ 44 Figure 83 shows an energy-level diagram for linear N_2H_2. How many bonds are there in this molecule? Does this agree with the simple Lewis structure for N_2H_2?

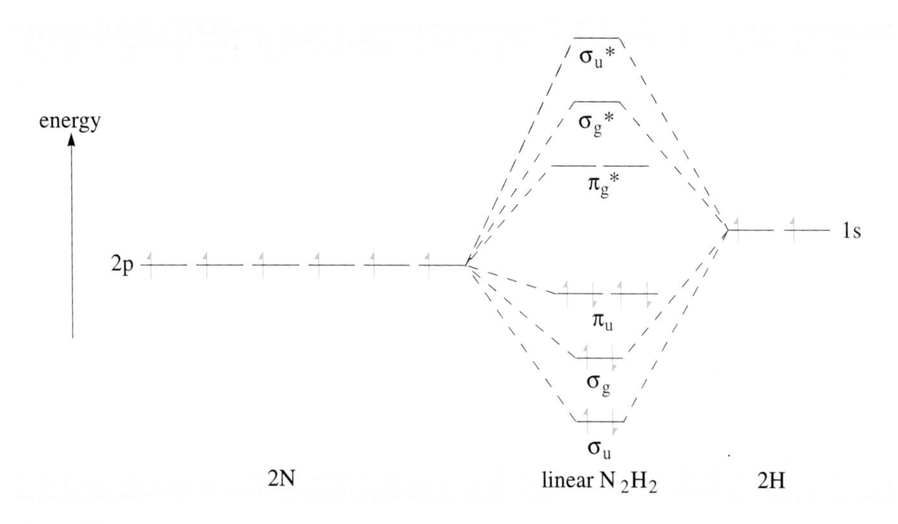

Figure 83 Orbital energy-level diagram for N_2H_2.

SAQ 45 Figure 84 shows the molecule OF_2 and a set of axes defining x, y and z. Determine the representation of two $2p_y$ orbitals, one on each fluorine. Will they combine with any orbitals on oxygen? If so, describe the resulting combination as σ or π, bonding, non-bonding or antibonding. Sketch the combination(s). Sketch the combination of $2p_y$ on fluorine that does not combine with the 2p orbitals on oxygen if there is one. Describe this orbital in terms of bonds or lone pairs.

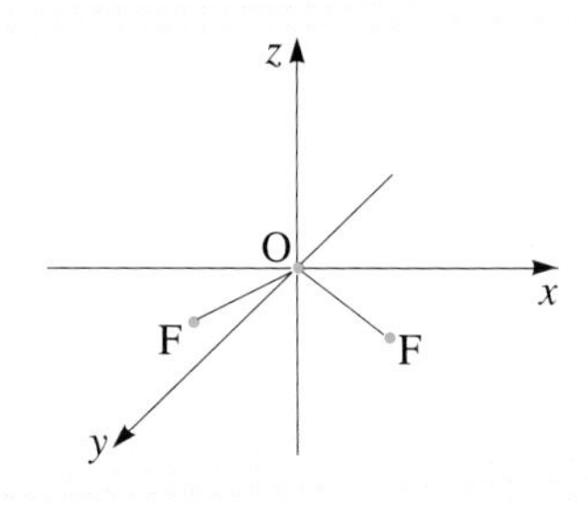

Figure 84

OBJECTIVES FOR BLOCK 4

Now that you have completed Block 4 you should be able to do the following things:

1 Recognize valid definitions of and use in a correct context the terms, concepts and principles in the following Table.

List of scientific terms, concepts and principles used in Block 4

Term	Page no	Term	Page no
antibonding orbital	46	nodal plane	45
axial group (site)	19	non-bonding orbital	70
axis of symmetry	25	octet	6
bonding orbital	43	odd-electron molecule	21
bond order	54	orbital	41
boundary surface	48	paramagnetism	56
centre of symmetry	30	π orbital	52

List of scientific terms, concepts and principles used in Block 4 - continued

2 Draw a Lewis structure for a given covalent molecule or ion. (Information as to which atoms are joined together will be given to you if there is any possible ambiguity.) (SAQs 1–5)

3 Use valence-shell electron-pair repulsion theory to predict the shape of given covalent molecules. (SAQs 6–9)

4 Recognize the presence of planes, axes and centres of symmetry in a given molecule or other object. (SAQs 10–24)

5 Label the symmetry elements of a given molecule or other object using the symbols in Section 5.4 of this Block. (SAQs 17–19)

6 Given a flow chart such as the one on the fold-out page at the end of this Block, assign a molecule or other object to its symmetry point group. (SAQs 20–24)

7 Draw rough sketches of representations of atomic orbitals for 1s, 2s and 2p orbitals and of molecular orbitals formed from these atomic orbitals. (SAQ 45)

8 Draw simple orbital energy-level diagrams for diatomic molecules, and simple polyatomic molecules, given the energies of appropriate atomic orbitals. (SAQs 25, 26, 28–33, 42, 43)

9 Use orbital energy-level diagrams for simple molecules to predict properties of these molecules such as stability relative to the separate atoms, bond order, paramagnetism. (SAQs 27, 28–33, 44)

10 Given character tables, determine the irreducible representation(s) to which given atomic orbitals in a molecule belong. You will be expected to deal only with molecules belonging to the point groups $\mathbf{C}_1$, $\mathbf{C}_s$, $\mathbf{C}_i$, $\mathbf{C}_2$, $\mathbf{C}_{2v}$, $\mathbf{C}_{2h}$ and $\mathbf{D}_{2h}$ (SAQs 35–42)

11 Use the results from simple polyatomic molecules to understand the bonding in other molecules. (SAQs 34, 45)

SAQ ANSWERS AND COMMENTS

SAQ 1 (*Objective 2*)

(a)

F N F
F

(b)

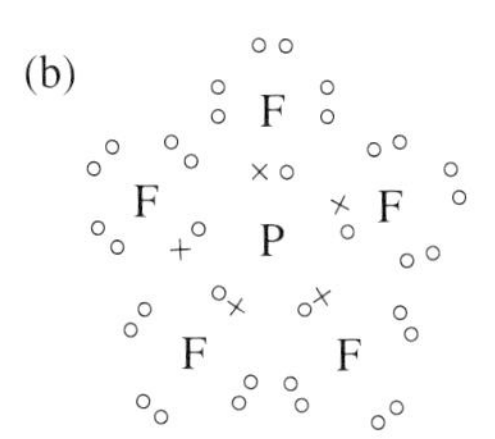

(c)

F
F S F
F

(d)

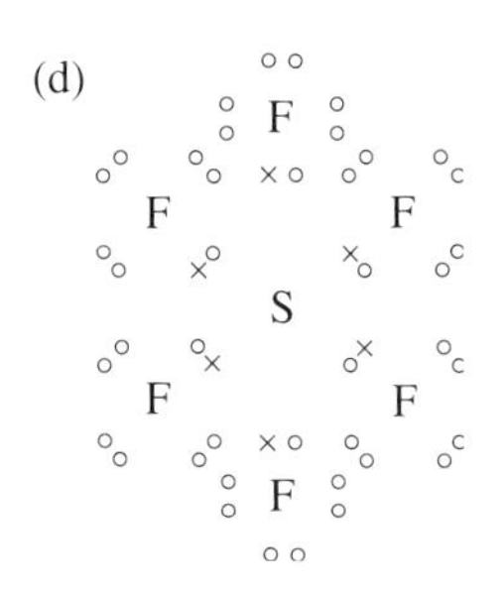

SAQ 2 (*Objective 2*)

(a)

H O Cl

(b)

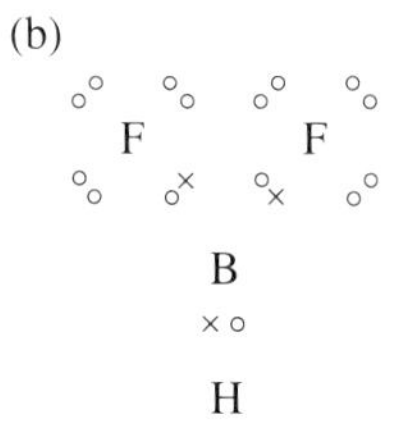

(c)

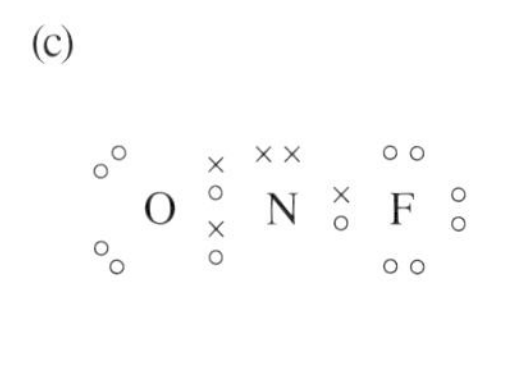

(d)

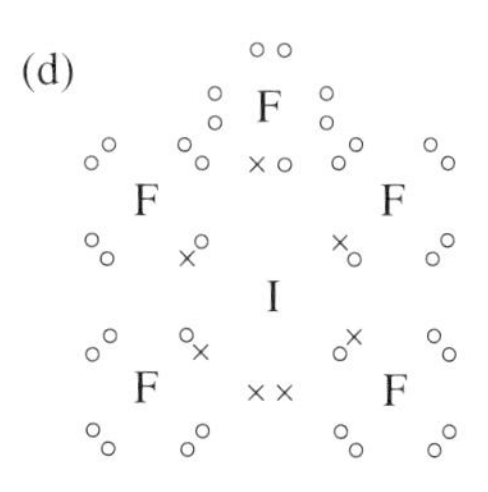

SAQ 3 (*Objective 2*)

(a) H—O—Cl:

(b)

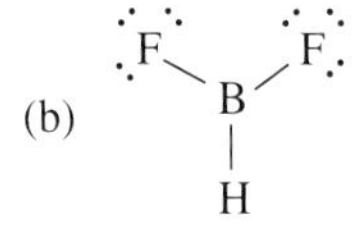

(c) O=N—F:

(d)

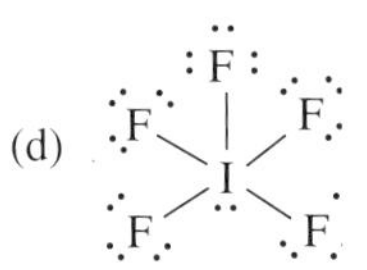

SAQ 4 (*Objective 2*)

The Lewis structure for Al_2Cl_6 is

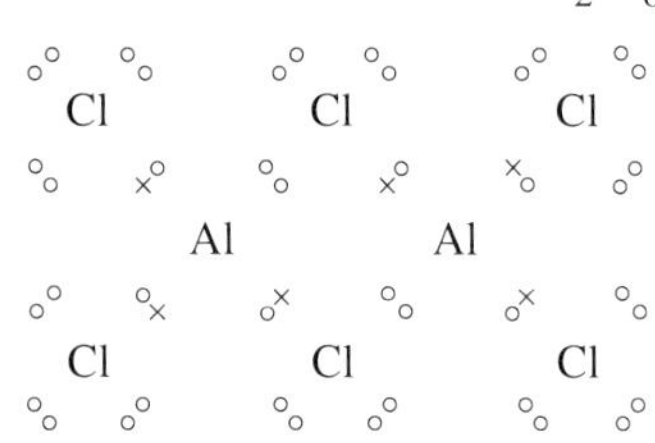

Aluminium has only three valence electrons and completes its octet by forming a dative bond with one of the chlorines in the middle. The structural formula is thus

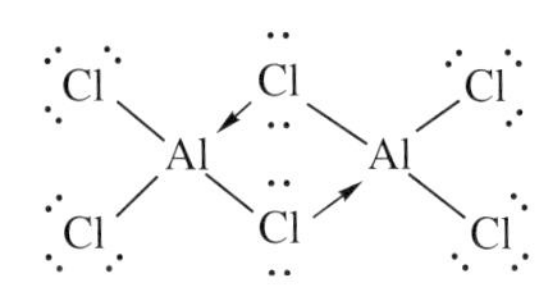

SAQ 5 (*Objective 2*)

The Lewis structures and structural formulae are as follows.

(a) H C O H

H\C=O/H

(b) H N N H

H—N=N—H

(c) Sulphur, being in the third row of the Periodic Table, can expand its octet, in this case to twelve electrons:

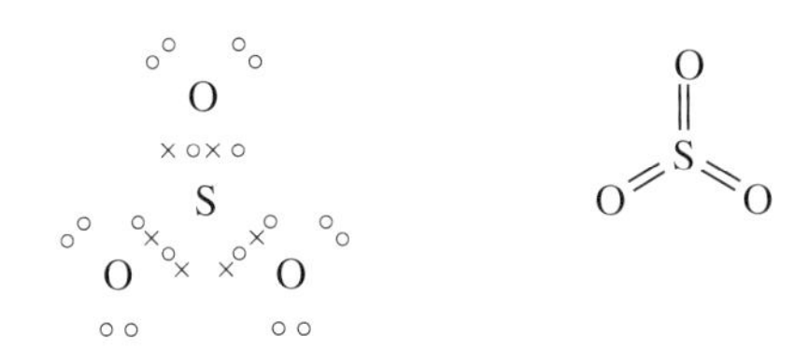

O=S(=O)=O

(d)

H × P × H (dot-and-cross), with H below P

H–P–H, with a lone pair on P and a third H below P

(e) Phosphorus has five valence electrons and boron has three, so in this adduct the lone pair on phosphorus is donated to boron, forming a dative bond:

H ∘ P ∘ B × H (dot-and-cross), with two H on P (one above, one below) and two H on B (one above, one below)

$H_3P \rightarrow BH_3$

(f) Nitrogen cannot expand its octet and so must form one double bond and one dative bond, but we do not expect the terminal N—O bonds to differ in length and so we must use resonance structures:

O, O, N × O × N, O, O (dot-and-cross)

O–N–O–N–O with O on each N: four resonance structures, linked by ⟷

SAQ 6 (*Objective 3*) (a) HOCl has the structural formula shown in the margin. There are four repulsion axes – two single bonds and two lone pairs. Four repulsion axes form a tetrahedral shape but lone pairs take up more room than single bonds and so HOCl will be bent with an H—O—Cl angle of less than 109.5°.

H—Ö—Cl (two lone pairs on O)

(b) HBF_2 has three single bonds around boron (see answer to SAQ 3). These form three repulsion axes. Three repulsion axes form a triangular shape and so HBF_2 will be planar. The angles in the regular shape are 120°. It is probable that B—F bonds take up more room than B–H and so the F—B—F angle would be greater than 120° and the H—B—F angles less than 120°.

(c) The structural formula of CO_2 is O=C=O. There are two repulsion axes (the two double bonds) and so the molecule will be linear.

(d) The structural formula of HCN is H—C≡N. Again there are two repulsion axes and the molecule is linear.

(e) SO_2F_2 has the structural formula shown in the margin; it has four repulsion axes. The molecule will be approximately tetrahedral. The S=O bonds will take up more room than S—F, and so the O—S—O angle will be greater than 109.5° and the F—S—F angle less than 109.5°.

SAQ 7 *(Objective 3)* (a) The structural formula of ICl_2^- is

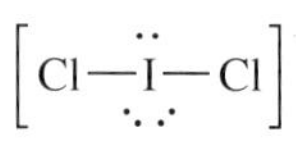

$[Cl{-}\ddot{I}{-}Cl]^-$

There are five repulsion axes. These will form a trigonal bipyramid. The lone pairs take up more room than the single bonds and so will occupy the equatorial positions. Thus the I—Cl bonds will occupy the axial positions and the ion will be linear.

(b) $[PCl_6]^-$

Cl
Cl Cl
P
Cl Cl
Cl

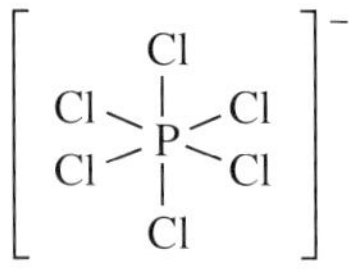

There are six repulsion axes and the ion is octahedral.

(c) $[NO_3]^-$

O O
N
O

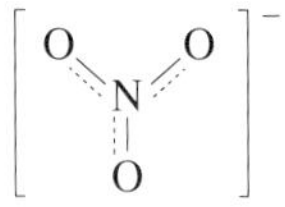

There are three repulsion axes and the ion is trigonal planar.

(d) The structural formula of NO_2^+ is $[O{=}N{=}O]^+$. There are two repulsion axes and the ion is linear. It is in fact isoelectronic with (has the same number of electrons as) CO_2, and so it is not surprising that it has the same shape.

SAQ 8 *(Objective 3)* (i) a; (ii) c; (iii) a; (iv) a; (v) a; (vi) a; (vii) c. In i, iii, iv, v and vi there are no lone pairs on the central atom, and so these molecules and ions are tetrahedral. In ii and vii there are five repulsion axes, one of which is a lone pair, and so the molecules are the same shape as SF_4.

SAQ 9 *(Objective 3)* (a) The structural formula of hydrogen peroxide is shown in the margin. Each oxygen has four repulsion axes around it: two lone pairs and two single bonds. These will be arranged approximately tetrahedrally, but because lone pairs take up more room than single bonds, the O—O—H angles will be less than 109.5°. This agrees with the structure shown. VSEPR does not, however, tell us anything about the relative orientations of the two O—H bonds; the molecule could be planar or non-planar, and the two hydrogens could be on the same side or opposite sides of the O—O bond.

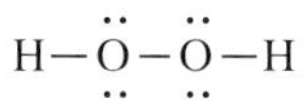

H—Ö—Ö—H

(b) The structural formula of N_2F_4 is

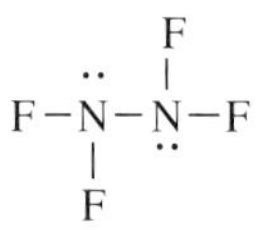

F
F—N—N—F
F

Each nitrogen thus has four repulsion axes around it: three single bonds and a lone pair. These will form a roughly tetrahedral arrangement around the nitrogen, with the lone pair occupying more space than the bonds so that the F—N—F angle will be less than 109.5°. This is as observed. Again, however, VSEPR cannot distinguish different orientations of the two NF_2 groups so that, on this theory, the shape

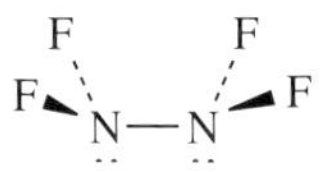

F F
F N—N F

for example, is as likely as the one shown in the question.

(c) The structural formula of $[SO_3NH_2]^-$ is

```
[     O          ]-          [     O          ]-          [     O          ]-
[     ↑   ..     ]           [     ‖   ..     ]           [     ↑   ..     ]
[ O═S—N—H        ]   ←→      [ O←S—N—H        ]   ←→      [ O←S—N—H        ]
[     ↓   |      ]           [     ↓   |      ]           [     ‖   |      ]
[     O   H      ]           [     O   H      ]           [     O   H      ]
```

The arrangement around both S and N is approximately tetrahedral because there are four repulsion axes around each. The S—O bonds would be expected to occupy more space than the S—N bond, and so the observed angles of greater than 109.5° for O—S—O and less than 109.5° for N—S—O are in agreement with VSEPR. However, N—H and S—N bonds would be expected to occupy less space than a lone pair and so the S—N—H and H—N—H angles would be predicted to be less than 109.5°. The observed angle is slightly greater.

SAQ 10 (*Objective 4*) (a) BrF_5 is square-based pyramidal. The axial Br—F bond is a fourfold axis and this is the only axis (Figure 85).

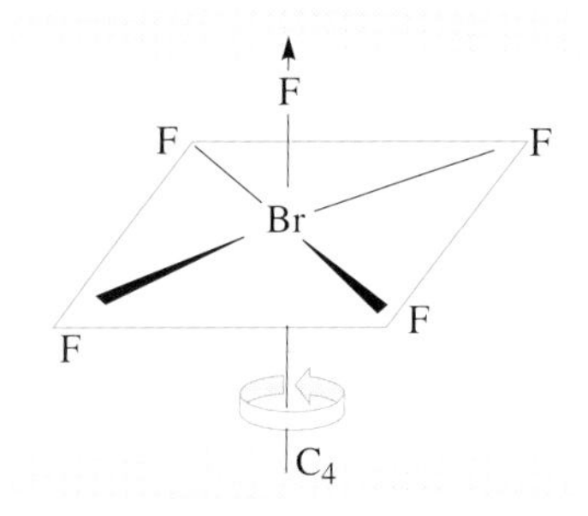

Figure 85 The fourfold axis in BrF_5.

(b) SO_2 is bent and has a twofold axis like that in OF_2.

(c) $POCl_3$ is pyramidal and the P—O bond forms a threefold axis.

(d) NO_3^- is trigonal planar. There is a threefold axis perpendicular to the plane of the molecule and each N—O bond is a twofold axis.

(e) HBF_2 is planar. The B—H bond is a twofold axis.

SAQ 11 (*Objective 4*) (a) Threefold; (b) fourfold; (c) twofold; (d) fourfold.

SAQ 12 (*Objective 4*) (a) The three equatorial P—F bonds lie in a plane of symmetry and there are three planes of symmetry each containing both axial P—F bonds and one equatorial P—F bond, making a total of four planes of symmetry. PF_5 is discussed on video sequence 5.

(b) The two equatorial S—F bonds lie in a plane of symmetry, and there is another plane of symmetry that contains the two axial S—F bonds and bisects the equatorial F—S—F angle. Thus SF_4 has two planes of symmetry (see Figure 26).

(c) The five equatorial I—F bonds form a plane and there are five planes containing two axial I—F bonds and one equatorial I—F bond, giving a total of six planes.

SAQ 13 (*Objective 4*) (i) c; (ii) b; (iii) a; (iv) b; (v) a; (vi) d; (vii) b. The plane of a molecule is always a symmetry plane. In i it is neither vertical nor horizontal because HOF has no axis of symmetry. In ii, the C_2 axis is at right-angles to the molecular plane and so the plane is horizontal. In v, the plane contains the ∞-fold axis and so is vertical. In vii, there is a threefold axis at right-angles to the plane and three twofold axes in the plane. The threefold axis is the principal axis and so the plane is horizontal. The plane in iii contains the P=O bond, which is the principal (threefold) axis and so is vertical. The molecular axis of CO_2 is the principal ∞-fold axis and so the plane is horizontal. Reflection through the plane of the three F atoms in NF_3 would bring the N atom through to the other side, and so this is *not* a plane of symmetry.

SAQ 14 (*Objective 4*) (a) No; (b) yes; (c) no; (d) yes; (e) yes.

SAQ 15 (*Objective 4*) Bent triatomics do not have a centre of symmetry. Only linear triatomics with identical outer atoms will have such a symmetry element. The molecule must therefore be straight: Cl—Ca—Cl.

SAQ 16 (*Objective 4*) Neither tetrahedral nor SF_4-shaped molecules have centres of symmetry. Thus $[PtCl_4]^{2-}$ must be square planar.

SAQ 17 (*Objectives 4 and 5*) (a) BF_3 has four planes of symmetry (one horizontal), one threefold axis and three twofold axes. The symbol for a threefold axis is C_3 and that for a twofold axis is C_2. Our list is thus $C_3 + 3C_2 + 3\sigma_v + \sigma_h$.

(b) SF_4 has only one twofold axis and two planes of symmetry, both vertical. The list for SF_4 is thus $C_2 + 2\sigma_v$.

(c) IF_7 has a fivefold axis (along the I—F axial bonds), five twofold axes (along equatorial I—F bonds) and six planes of symmetry (one horizontal). The complete list is thus $C_5 + 5C_2 + 5\sigma_v + \sigma_h$.

SAQ 18 (*Objectives 4 and 5*) The environment of Pb^{2+} in PbO displays the same symmetry as the BrF_5 molecule. There are a fourfold axis and four planes of symmetry, all vertical (Figure 86). Our list is thus $C_4 + 4\sigma_v$.

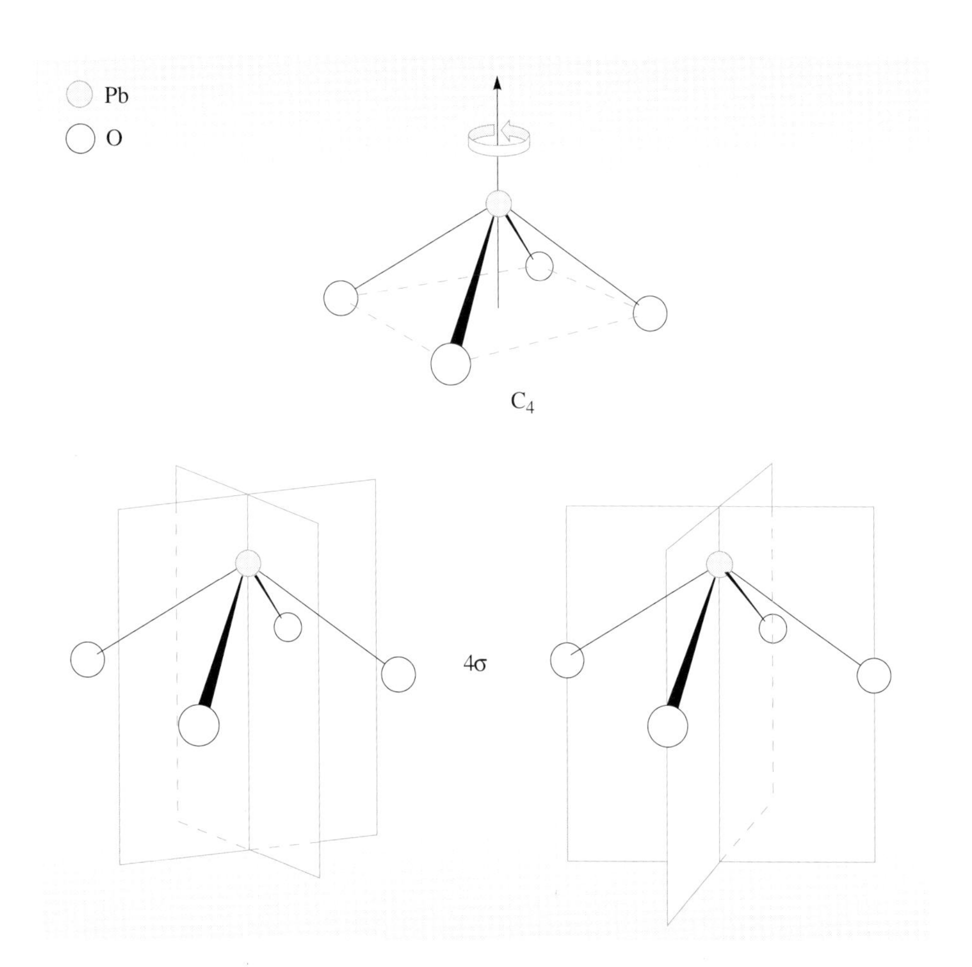

Figure 86 Symmetry elements in crystalline PbO.

SAQ 19 (*Objectives 4 and 5*)

(a) A spoon has only a plane of symmetry, σ.

(b) $C_7 + 7C_2 + 7\sigma_v + \sigma_h$.

(c) $C_\infty + \infty\sigma_v$.

(d) $C_6 + 6C_2 + 6\sigma_v + \sigma_h + i$.

(e) A hand has no symmetry elements.

SAQ 20 (*Objectives 4 and 6*) (a) SO_2 is neither tetrahedral nor octahedral. It has one C_2 axis. The largest value of *n* is thus 2, and there are not two C_2 axes perpendicular to the C_2 axes. There are two planes of symmetry containing the C_2 axis. So SO_2 belongs to the point group $\mathbf{C_{2v}}$.

(b) KrF_2 is linear. It is therefore neither tetrahedral nor octahedral. The molecular axis is a C_∞ axis and ∞ must be the highest value of n, since you cannot get a number larger than infinity! Are there an infinite number of C_2 axes perpendicular to the C_∞ axis? Yes, any line through the Kr atom perpendicular to the molecular axis is a C_2 axis for KrF_2, and there are an infinite number of such lines. Is there a plane of symmetry perpendicular to the C_∞ axis? Yes, the plane through the krypton atom at right-angles to the molecular axis is a plane of symmetry. KrF_2 therefore belongs to the group $\mathbf{D}_{\infty h}$.

(c) The molecular axis of CO is a C_∞ axis but there is no C_2 axis perpendicular to it. There are an infinite number of planes of symmetry containing the C_∞ axis. It therefore belongs to $\mathbf{C}_{\infty v}$. You may remember that we said all heteronuclear diatomic molecules belong to this point group.

(d) $POCl_3$ is only approximately tetrahedral and, because one of the atoms surrounding phosphorus is different from the other three, it does not belong to the group $\mathbf{T}_d$. It is not octahedral. It does have a C_3 axis (the P=O bond), but it has no other axes of symmetry. It does have three planes of symmetry containing the C_3 axis. (These each contain the P=O bond and one P—Cl bond.) $POCl_3$ thus belongs to the group $\mathbf{C}_{3v}$.

(e) NSF is obviously neither tetrahedral nor octahedral. It does not have an axis of symmetry. It does, however, have a plane of symmetry, so it belongs to the group $\mathbf{C}_s$.

It is interesting to note that all triatomic molecules must have at least one element of symmetry. This is because three atoms have to be in the same plane and this plane will be a plane of symmetry. Molecules with a plane of symmetry cannot be optically active because they are identical with their mirror images. Hence all triatomic molecules are non-chiral.

SAQ 21 *(Objectives 4 and 6)* (a) A saucer belongs to $\mathbf{C}_{\infty v}$. It has a C_∞ axis through the centre of the saucer, and an infinite number of vertical planes of symmetry.

(b) A spoon has only one symmetry element – a plane of symmetry. It therefore belongs to $\mathbf{C}_s$.

(c) A snowflake has a C_6 axis, six C_2 axes at right-angles to this axis, seven planes of symmetry, one of which is perpendicular to the C_6 axis, and a centre of symmetry. Using the flow chart, we can see that the snowflake belongs to the point group $\mathbf{D}_{6h}$.

(d) A starfish has a fivefold axis, but no C_2 axes perpendicular to it. There are five vertical planes of symmetry. It therefore belongs to the point group $\mathbf{C}_{5v}$.

SAQ 22 *(Objectives 4 and 6)* (a) Group (v): all homonuclear diatomic molecules belong to $\mathbf{D}_{\infty h}$.

(b) Group (iv): ethene has three C_2 axes, which are all perpendicular to each other, and so can be counted as $C_2 + 2C_2$ perpendicular to the first.

(c) Group (ii): see $POCl_3$ which has the same symmetry.

(d) Group (i): no symmetry operation swaps the central atom for one of the outer two, so the fact that the central atom here is identical with the outer two atoms makes no difference to the symmetry. O_3 thus belongs to the same point group as SO_2.

(e) Group (iii): all heteronuclear diatomic molecules belong to $\mathbf{C}_{\infty v}$.

SAQ 23 *(Objectives 4 and 6)* None of the groups $\mathbf{C}_{nv}$, and only those groups $\mathbf{D}_{nh}$ for which n is even, contain a centre of symmetry. $\mathbf{C}_i$ contains a centre of symmetry by definition. So the groups containing molecules with a centre of symmetry are $\mathbf{C}_i$ and $\mathbf{D}_{2h}$.

SAQ 24 *(Objectives 4 and 6)* In CF_4, each C—F bond is a threefold axis so there are four altogether. The C—Cl bond in CF_3Cl is a threefold axis but this is the only one. CF_2Cl_2 has no threefold axes.

CF_4 belongs to $\mathbf{T}_d$, CF_3Cl to $\mathbf{C}_{3v}$ and CF_2Cl_2 to $\mathbf{C}_{2v}$.

SAQ 25 (*Objective 8*) You should have obtained a diagram like Figure 87 for He_2. There are two electrons in bonding orbitals and two in antibonding orbitals. The antibonding orbital is slightly higher in energy compared with the 1s, than the bonding orbital is lower, so that the He_2 molecule will have a higher electron energy than two He atoms. Consequently, we would expect He atoms to be favoured over He_2.

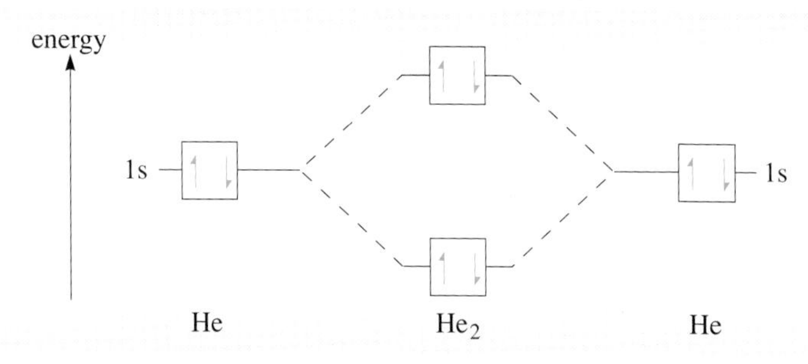

Figure 87 Orbital energy-level diagram for He_2.

SAQ 26 (*Objective 8*) Be_2 will have eight electrons, since Be has four ($1s^22s^2$). The orbital energy-level diagram for Be_2 is given in Figure 88. The electrons are fed into the orbitals, two into the $1\sigma_g$, two into the $1\sigma_u$, two into the $2\sigma_g$ and two into the $2\sigma_u$.

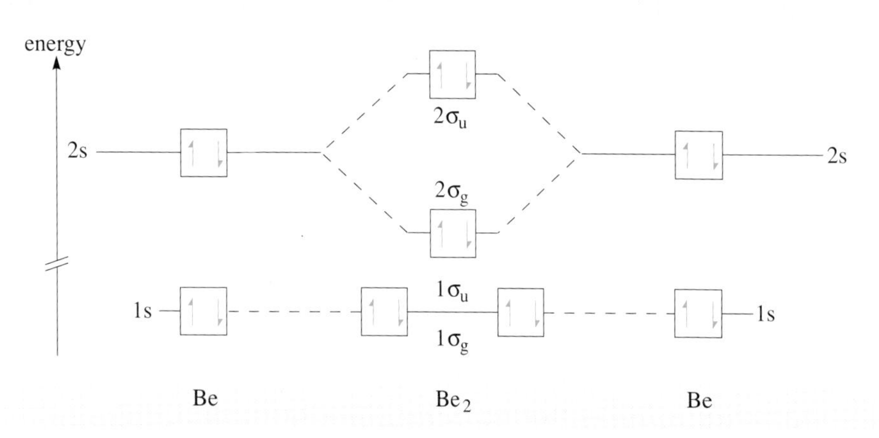

Figure 88 Orbital energy-level diagram for Be_2.

You probably predicted that Be_2 would not exist. At room temperature and atmospheric pressure, beryllium is a highly poisonous grey metal. Be_2 is not found at room temperature, and the observed spectra were obtained by trapping Be atoms in a solid matrix at low temperature. Where pairs of Be atoms were trapped together, they gave spectra characteristic of Be_2. It is probably more stable than predicted from Figure 88 because there is some involvement of 2p orbitals in the bonding.

SAQ 27 (*Objective 9*) (a) 1; (b) 0; (c) 1. H_2 has one pair of electrons in a bonding orbital and none in antibonding. He_2 has one pair in bonding and one in antibonding, and Li_2 has two pairs in bonding and one pair in antibonding. Note that molecules with bond orders of 0, like He_2, are those that we predict not to exist.

SAQ 28 (*Objectives 8 and 9*) Figure 89 shows the orbital energy-level diagram for Ne_2. You should have predicted that it would not exist, and indeed it has never been observed.

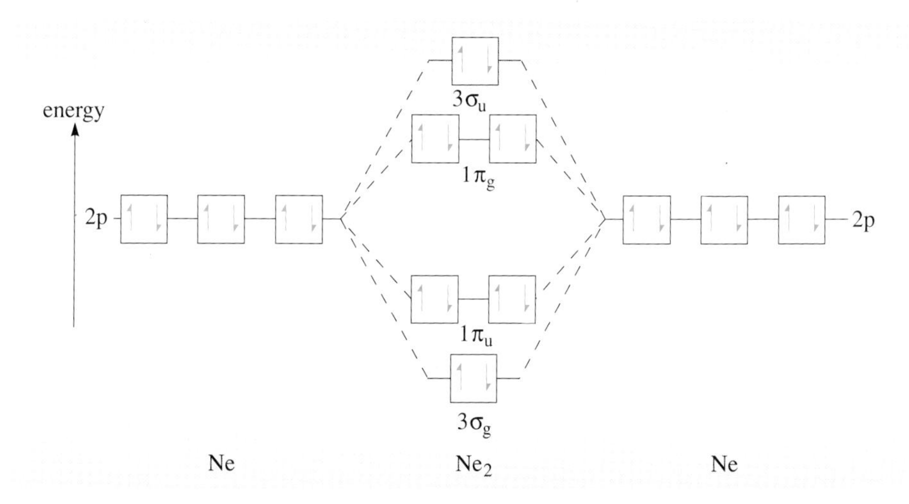

Figure 89 Orbital energy-level diagram for Ne_2.

SAQ 29 (*Objective 9*) The orbital energy-level diagram for O_2^+ is the same as that for O_2 except that there is one less electron to feed in (Figure 90). As there is now only one electron in the antibonding orbital $1\pi_g$, the bond order in O_2^+ is $2\frac{1}{2}$ compared with a bond order of 2 in O_2. We would therefore expect the bond length of O_2^+ to be shorter than that in O_2 and the dissociation energy to be larger.

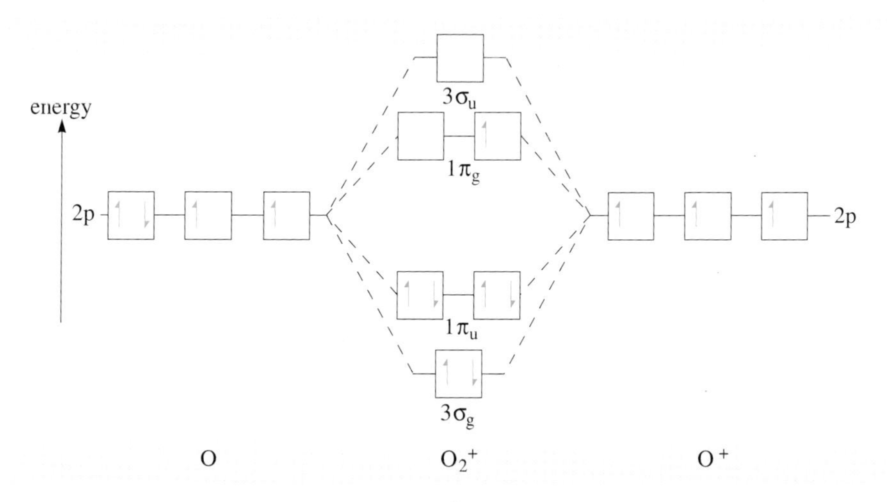

Figure 90 Orbital energy-level diagram for O_2^+.

The observed bond lengths of O_2 and O_2^+ are 121 pm and 112 pm, respectively, and the bond dissociation energies are 498 kJ mol^{-1} and 623 kJ mol^{-1}, respectively. The electron that has to be removed from O_2 to form O_2^+ is antibonding, and we might expect this to be fairly easily removed. This is borne out by the existence of salts of O_2^+, such as $O_2^+[PtF_6]^-$ and $O_2^+[PF_6]^-$. Since O_2 has two electrons in antibonding orbitals, you may have thought that we could also form O_2^{2+} salts, in which the bond order would be 3. The energy required to remove two electrons from O_2, however, is so large that we would need a very large lattice energy to be able to form stable salts.

SAQ 30 (*Objective 9*) N_2^+ has one less electron than N_2. The electron lost will come from the $1\pi_u$ bonding orbital, thus reducing the bond order by $\frac{1}{2}$ to $2\frac{1}{2}$. We would therefore expect N_2^+ to have a smaller dissociation energy than N_2.

N_2^- has one more electron than N_2, but because this electron has to go into an antibonding orbital ($1\pi_g$) the bond order of N_2^- is also reduced to $2\frac{1}{2}$. We predict therefore that N_2^+ and N_2^- will have similar dissociation energies, which will be less than that of N_2.

SAQ 31 (*Objective 9*) The diagram is shown in Figure 91.

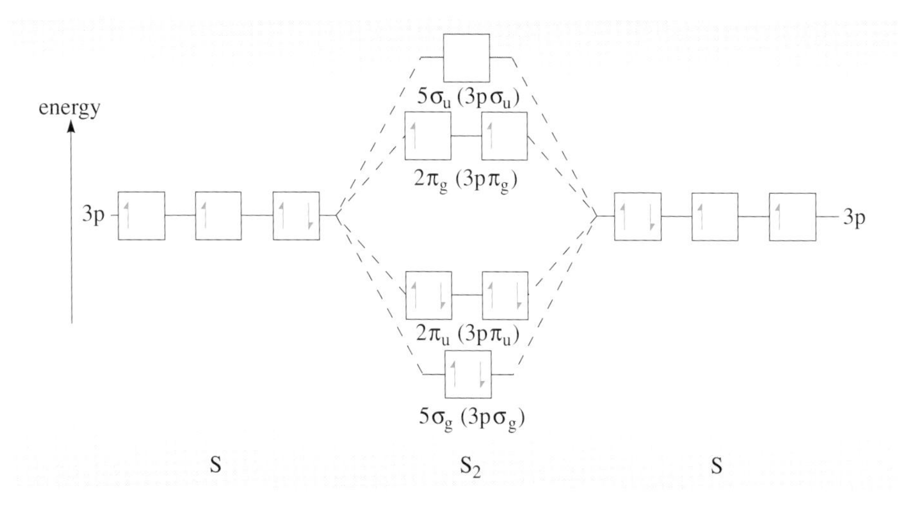

Figure 91 The orbital energy-level diagram for S_2.

This is almost identical with the diagram for O_2 (Figure 57), the only differences being that we use 3p rather than 2p orbitals and that the actual energies of all the orbitals will differ from those in O_2. The bond order is 2 and the molecule is paramagnetic.

SAQ 32 (*Objective 9*) Figure 92 shows an orbital energy-level diagram for OF. This resembles that for NO except that there are two more electrons. These two will go into the $2\pi^*$ antibonding orbital, giving OF a bond order of $1\frac{1}{2}$. OF is thus stable with respect to O and F atoms. It is not observed probably because of the disproportionation reaction:

$$2OF(g) = OF_2(g) + \tfrac{1}{2}O_2(g)$$

or because of its instability with respect to the normal forms of O and F, O_2 and F_2:

$$2OF(g) = O_2(g) + F_2(g)$$

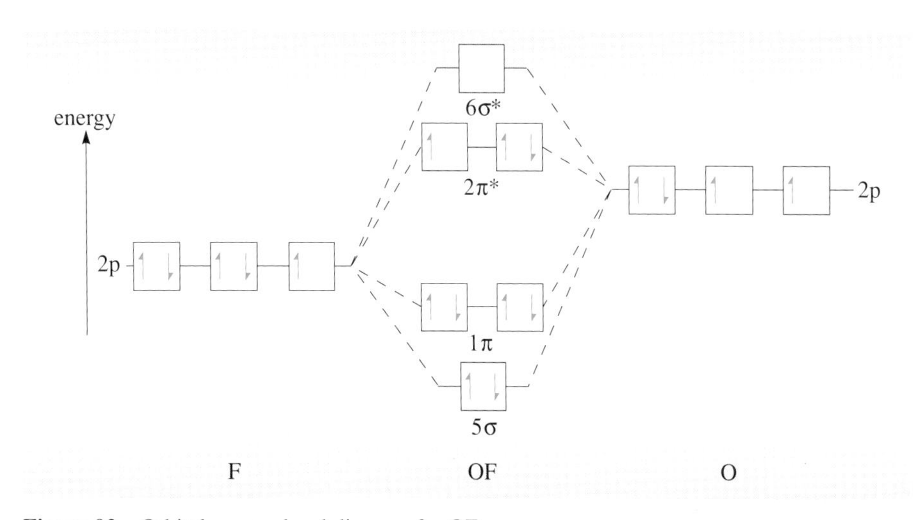

Figure 92 Orbital energy-level diagram for OF.

SAQ 33 (*Objectives 8 and 9*) The atomic orbitals nearest in energy for CN are C 2s and N 2p, but C 2s and C 2p are close in energy as are N 2s and N 2p. Thus you must also consider the C 2p and N 2s. Start by combining C 2p with N 2p and follow the method used for CO. The C 2p and N 2p combine to give 5σ, 1π, $2\pi^*$ and $6\sigma^*$. You now have to mix in the C 2s and N 2s. These will be of σ symmetry and so will combine with the nearest σ molecular orbital, 5σ. Mixing in C 2s and N 2s will raise the energy of 5σ, as this orbital now consists of N $2p_z$ combined with C $2p_z$ in a bonding manner and C 2s and N 2s in an antibonding manner. The 2s orbitals probably have a greater effect in CN than in CO, but, overall, 5σ is bonding. The orbital energy-level diagram is shown in Figure 93.

CN^- is isoelectronic with CO. CN^- has suitable orbitals to bind to iron and will do so if the energies of the CN^- and Fe orbitals are close. CN^- is a well-known poison that does in fact bind readily to many metal ions, including iron in haemoglobin.

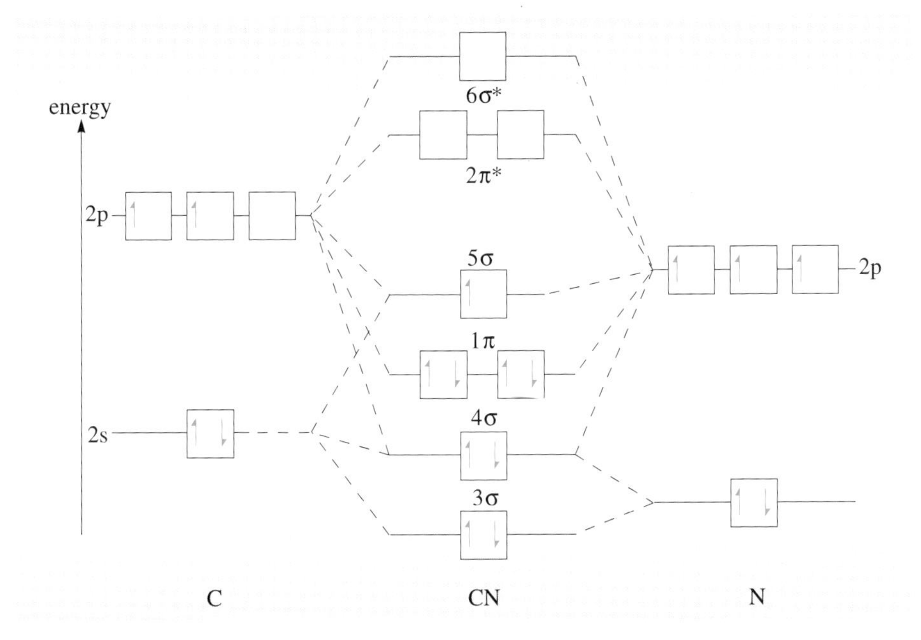

Figure 93 Orbital energy-level diagram for CN.

SAQ 34 (*Objective 11*) The orbital is above and below the molecular plane and so is a π orbital.

SAQ 35 (*Objective 10*) There are two orbitals and so under $\hat{I}$ we put 2. The $\hat{C}_2$ and $\hat{\sigma}_v(yz)$ operations swap the atoms and so get 0. This leaves $\hat{\sigma}_v(xz)$. The $2p_z$ orbitals lie in this plane and so are unchanged by reflection. Thus we get 2 0 2 0, which is $a_1 + b_1$.

SAQ 36 (*Objective 10*) Two orbitals give 2 under $\hat{I}$. $\hat{C}_2$ swaps the atoms and so gets 0. The definition of the x and y axes is open to choice. If we choose to put the x axis along the axial SF bonds (an equally valid choice is to make this the y axis) then $\hat{\sigma}_v(xz)$ swaps the two atoms and thus gets 0 and $\hat{\sigma}_v(yz)$ gets 2. This gives 2 0 0 2 or $a_1 + b_2$. The other choice of x and y axes gives $a_1 + b_1$. Either answer is valid, and it does not matter which you give so long as you state where your axes are. The equatorial and axial fluorines are considered separately because no symmetry operation interchanges them.

SAQ 37 (*Objective 10*) Suppose we label the hydrogen atoms 1, 2, 3 and 4 as in Figure 94. $\hat{C}_2(z)$ and $\hat{\sigma}_v(yz)$ interchange 1 with 4 and 2 with 3, $\hat{i}$ and $\hat{C}_2(y)$ interchange 1 with 3 and 2 with 4, and $\hat{C}_2(x)$ and $\hat{\sigma}_v(xy)$ interchange 1 with 2 and 3 with 4. Hence all four can be swapped.

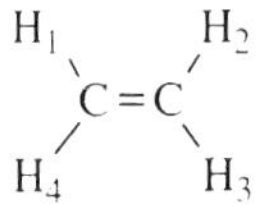

H_1 H_2
C=C
H_4 H_3

Figure 94 The ethene molecule.

All these operations that swap atoms get a 0 in the representation. This leaves only $\hat{I}$, which gets 4, and $\hat{\sigma}_v(xz)$ (the plane of the molecule), which also gets 4. The representation is thus 4 0 0 0 0 0 4 0, which is equivalent to $a_g + b_{2g} + b_{1u} + b_{3u}$.

SAQ 38 (*Objective 10*) The equatorial fluorines lie in the yz plane, so that the atoms are swapped by reflection through the xz plane and by rotation about the C_2 axis. There are two $2p_x$ orbitals, and so we put 2 under $\hat{I}$. We put 0 under $\hat{C}_2$ and $\hat{\sigma}_v(xz)$ because these operations swap the atoms. Finally, reflection in the yz plane swaps the lobes of $2p_x$, so that we have −2 under $\hat{\sigma}_v(yz)$. Thus the reducible representation of $2p_x$ on the equatorial fluorines is 2 0 0 −2, which is $a_2 + b_1$.

The $2p_y$ and $2p_z$ orbitals also have 2 under $\hat{I}$ and 0 under $\hat{C}_2$ and $\hat{\sigma}_v(xz)$. These orbitals lie in the yz plane, however, and so get +2 under $\hat{\sigma}_v(yz)$. Thus the reducible representation for $2p_y$ or $2p_z$ is 2 0 0 2. This is $a_1 + b_2$.

SAQ 39 (*Objective 10*) The valence orbitals of oxygen are 2s and 2p; the energy gap between them is large so we need only consider 2p. The two outer oxygens are interchanged by symmetry operations and so are equivalent. We thus consider the 2p orbitals on the central atom and then those on the two outer atoms together.

For the central atom (as in H_2O), $2p_x$ belongs to b_1, $2p_y$ to b_2 and $2p_z$ to a_1.

For the outer atoms we take first the two $2p_x$ orbitals. These have a representation $2\ 0\ 2\ 0 = a_1 + b_1$. The two $2p_z$ orbitals also give $2\ 0\ 2\ 0 = a_1 + b_1$. Finally, the two $2p_y$ orbitals give $2\ 0\ {-2}\ 0 = a_2 + b_2$.

So the $2p_x$ and $2p_z$ on the central atom will overlap with the $2p_x$ and $2p_z$ on the outer atoms to form six molecular orbitals altogether; these will be σ orbitals because the electron density in these orbitals lies in the molecular plane. $2p_y$ on the central atom combines with $2p_y$ on the outer atoms to give three π orbitals.

SAQ 40 (*Objective 10*) These will be the same as in H_2O because 4p has the same symmetry properties as 2p, that is a_1 bonding, b_1 bonding, a_1 antibonding, b_1 antibonding and b_2 non-bonding.

SAQ 41 (*Objective 10*) The representation of the 1s orbitals reduces to $a_g + b_{2g} + b_{1u} + b_{3u}$ (see SAQ 37). These will combine with $2p_x$ $(b_{2g} + b_{3u})$ and $2p_z$ $(a_g + b_{1u})$ on carbon and so these are the carbon orbitals that contribute to the C—H bond.

[Note that the C 2s orbitals belong to $(a_g + b_{1u})$ and, because the C 2s and C 2p are close in energy, the C—H (and C—C) σ bonds involve one s orbital and two p orbitals on each carbon. If we combine the 2s and two 2p on each carbon before combining them with each other and the hydrogen 1s, then we obtain the sp^2 hybrids that will be familiar to past or present students of S246.]

SAQ 42 (*Objectives 8 and 10*) The diagram is shown in Figure 95.

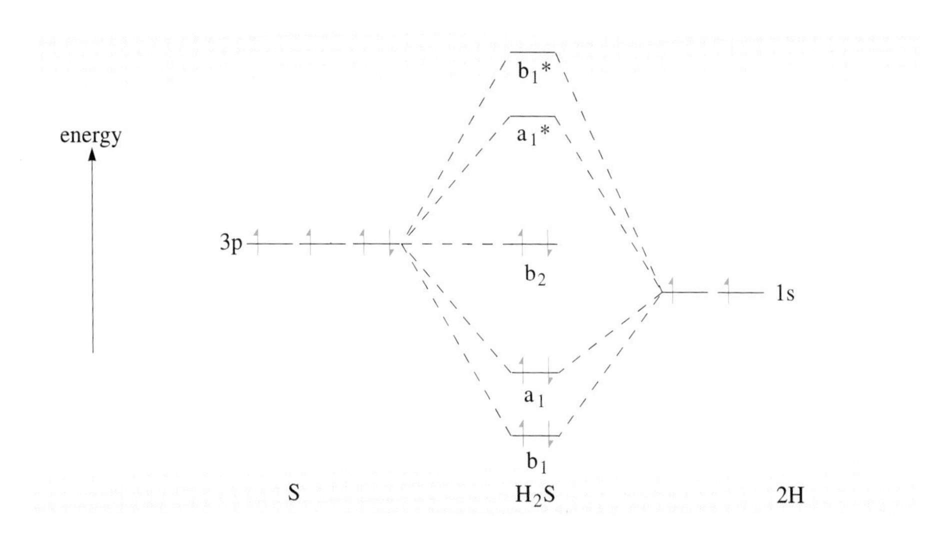

Figure 95 Orbital energy-level diagram for H_2S.

The $3d_{yz}$ on sulphur has the irreducible representation b_1. It will therefore combine with b_1 molecular orbitals. 3d is higher in energy than 3p in S, and so it will combine with the nearest b_1 orbital, which is the antibonding orbital. Two b_1 orbitals will result, and the lower one will be closer in energy to the antibonding a_1 than that shown in the Figure. If the 3d and 3p are close in energy there may be mixing in of $3d_{xz}$ to the b_1 bonding orbital. This will serve to lower the energy of this orbital thus strengthening the bonding.

SAQ 43 (*Objective 8*) In SAQ 39 you saw that the $2p_y$ orbital on the central atom combines with the $2p_y$ on the outer atom to give three π orbitals (Figure 96). The orbital in which all three $2p_y$ orbitals are in phase is a bonding orbital. That in which the phase alternates so that the central atom orbital is out of phase with the outer two is antibonding. Both of these orbitals will belong to b_2 because the $2p_y$ on the central atom does. The bonding orbital will lie below the O atom orbital and the antibonding orbital above. In both b_2 orbitals the orbitals on the two end atoms are in phase with each other. There is another π orbital in which these two are out of phase and there is no contribution from the central atom. This orbital is non-bonding, a_2.

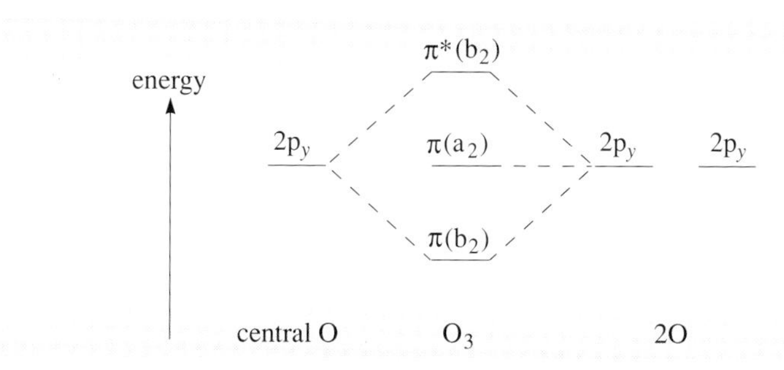

Figure 96 π orbital energy-level diagram for O_3.

The six σ orbitals (Figure 97) are formed from $2p_x$ (b_1) on the central atom combining with the b_1 combinations of $2p_x$ and $2p_z$ on the outer atoms, and from $2p_z$ (a_1) on the central atom combining with the a_1 combinations of $2p_x$ and $2p_z$ on the outer atoms. In each case (a_1 and b_1) we are combining three orbitals and so obtain three orbitals. These will be one bonding, one antibonding, and one non-bonding, so that there are two bonding, two antibonding and two non-bonding σ orbitals. The a_1 and b_1 bonding orbitals are unlikely to have identical energies, but we are not in a position to say which would be lower. We have therefore labelled the a_1 and b_1 bonding orbitals σ and the a_1 and b_1 antibonding orbitals σ*, and have not specified which is the lower energy.

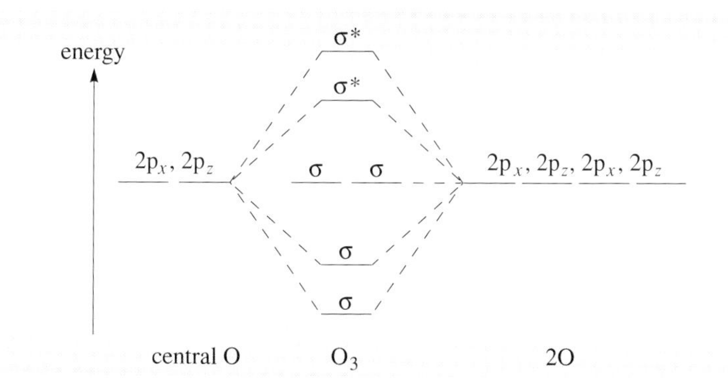

Figure 97 σ orbital energy-level diagram for O_3.

SAQ 44 (*Objective 9*) There are four bonds in this molecule, because there are eight electrons in bonding orbitals and none in antibonding. The simple Lewis structure and the structural formula are

H $\overset{\circ}{\times}$ N $\overset{\times}{\underset{\times}{\circ\circ}}$ N $\overset{\times}{\circ}$ H H—N̈=N̈—H

The 2s orbitals on nitrogen (not shown in Figure 83) would correspond to the lone pairs in the Lewis structure. The electrons in the σ bonding orbitals form the N—H bonds. The Lewis structure merely indicates a double bond, whereas the energy-level diagram indicates that there is a double π bond consisting of combinations of N 2p orbitals.

SAQ 45 *(Objectives 7 and 11)* There are two fluorines, and so there are two $2p_y$ orbitals and we put 2 under $\hat{I}$. The operations $\hat{C}_2$ and $\hat{\sigma}_v(yz)$ swap the two fluorines and so we put 0 under these operations. $\hat{\sigma}_v(xz)$ swaps positive and negative lobes on each $2p_y$ orbital and so we have –2 under this operation. The reducible representation is 2 0 –2 0, which is $a_2 + b_2$. The b_2 combination will combine with the $2p_y$ on oxygen. A sketch of the bonding combination is shown in Figure 98a and one of the antibonding combination in Figure 98b. These orbitals have electron density above and below the plane of the molecule and are thus π orbitals.

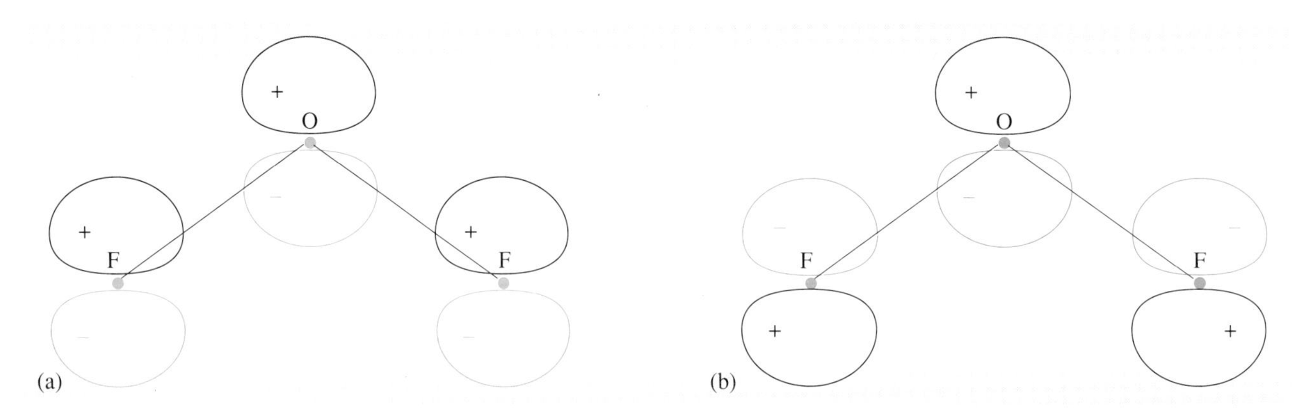

Figure 98 The b_2 orbitals in OF_2.

The a_2 combination is shown in Figure 99. This does not combine with any 2p orbitals on oxygen. It is non-bonding and will contribute to the lone pairs on the fluorines.

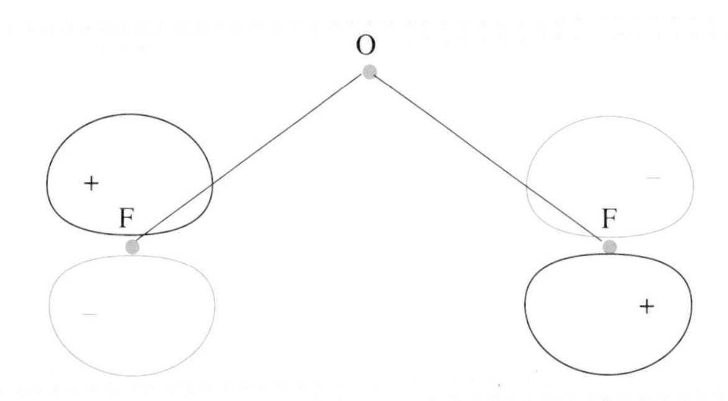

Figure 99 The a_2 orbital in OF_2.

ACKNOWLEDGEMENTS

Grateful acknowledgement is made to the following sources for permission to reproduce illustrations in this Block:

Figure 31 Brian Webster *Chemical Bonding Theory*, Blackwell Scientific Publications Ltd.

FLOW CHART FOR DETERMINING THE SYMMETRY POINT GROUP OF AN OBJECT

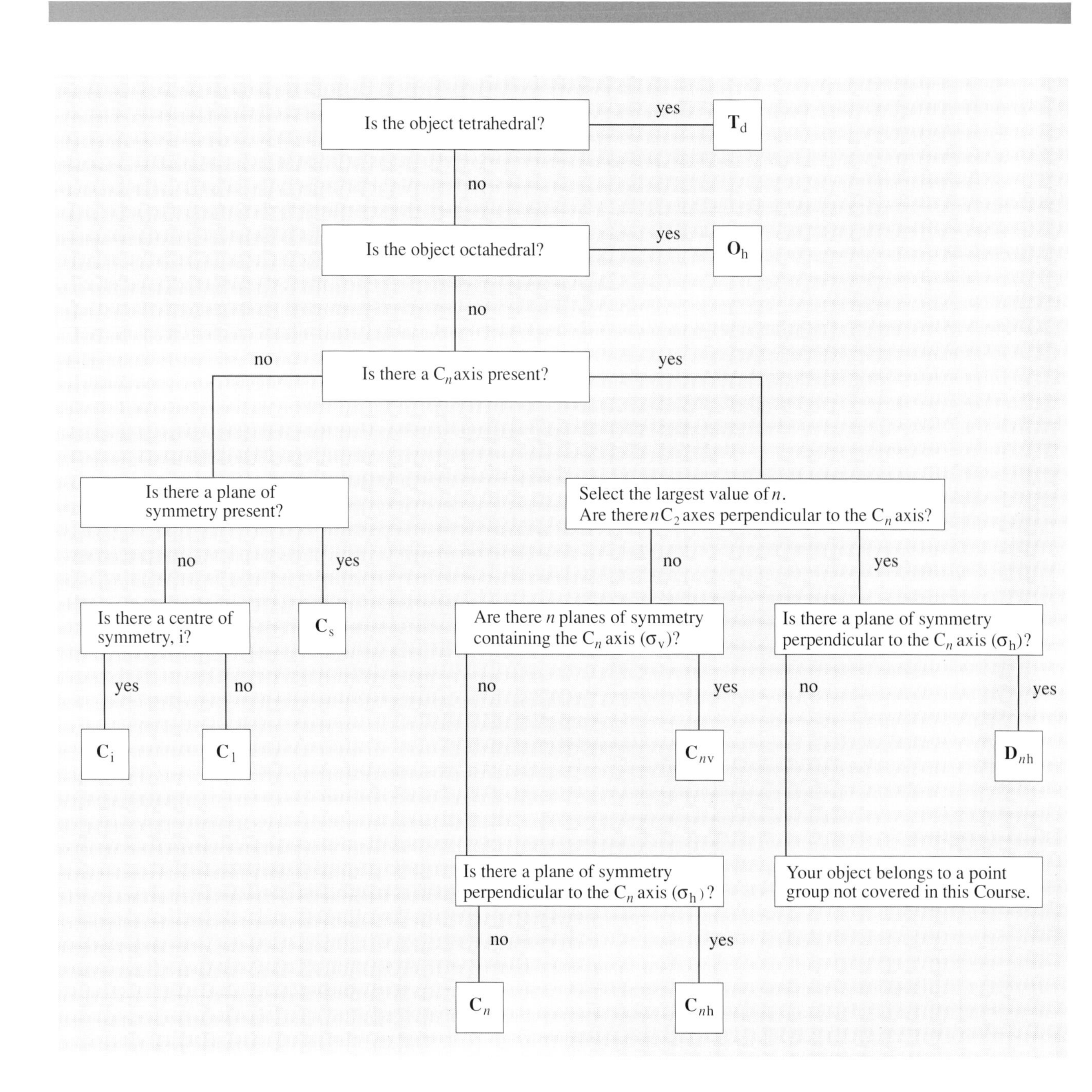

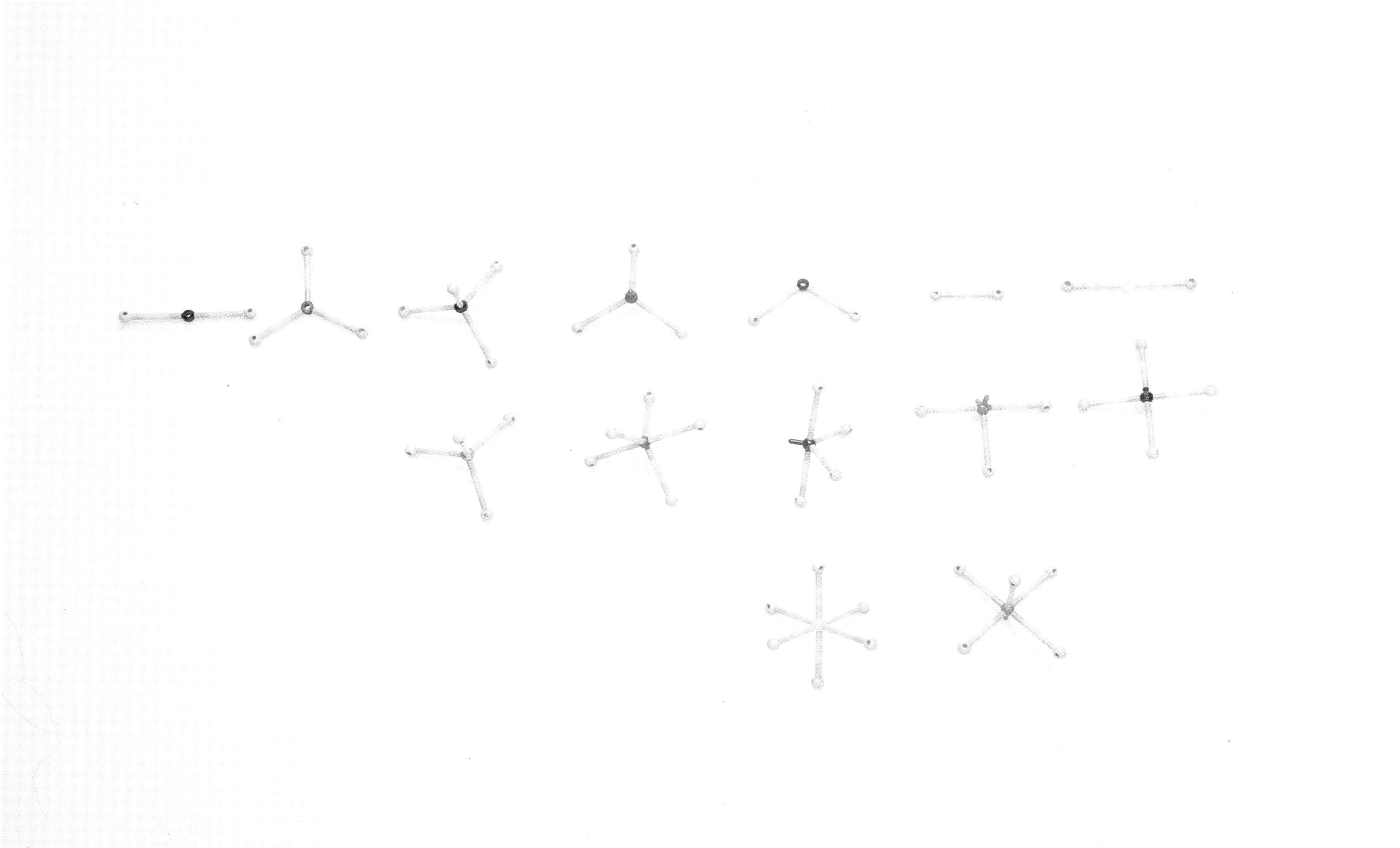

Plate 1 Orbit models of the fluorides discussed in Section 3.

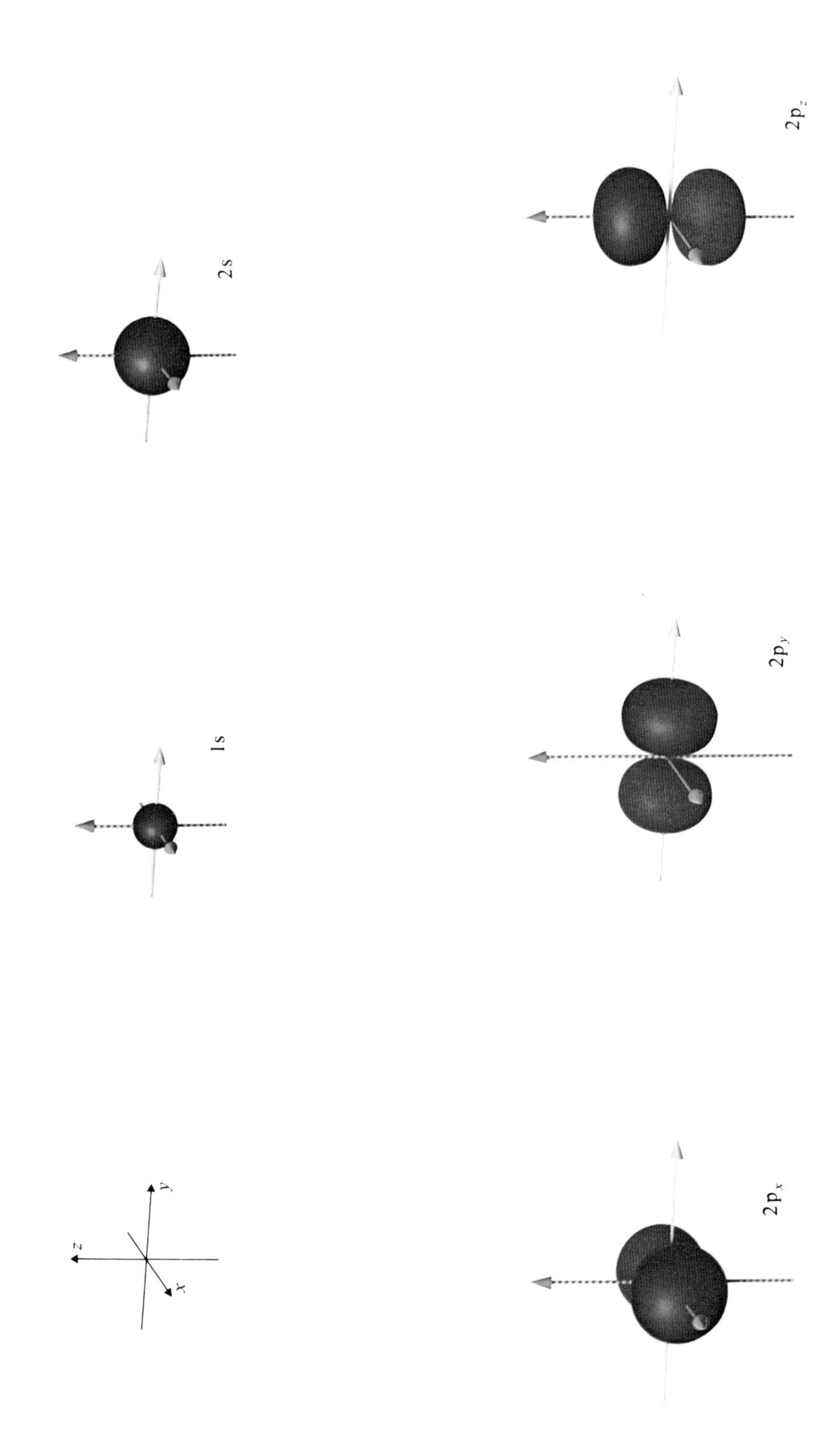

Plate 2 The shapes and relative sizes of 1s, 2s, 2p, and 3d orbitals for the hydrogen atom.

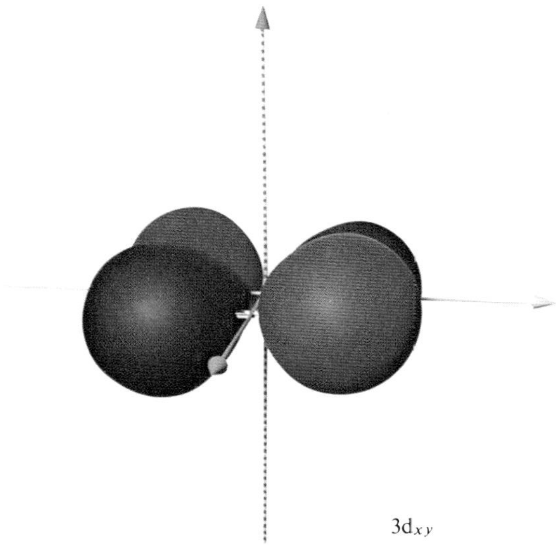

$3d_{xy}$  $3d_{xz}$ 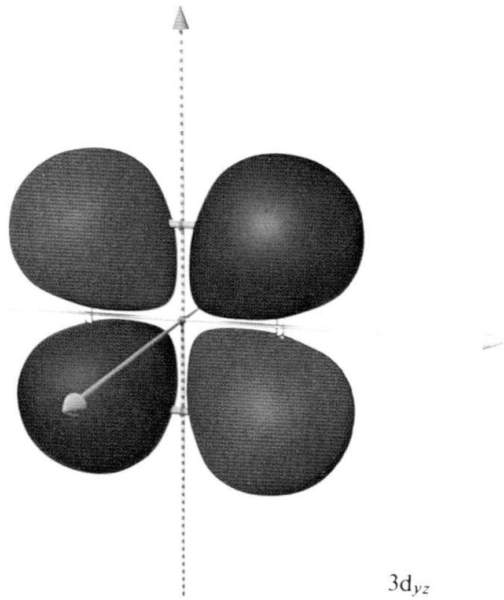 $3d_{yz}$

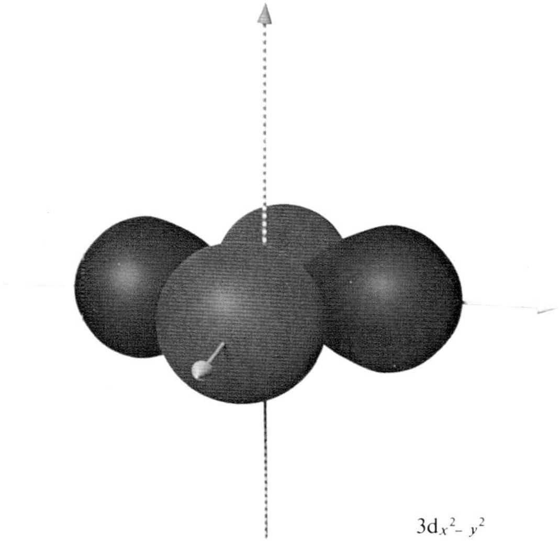

$3d_{x^2-y^2}$ 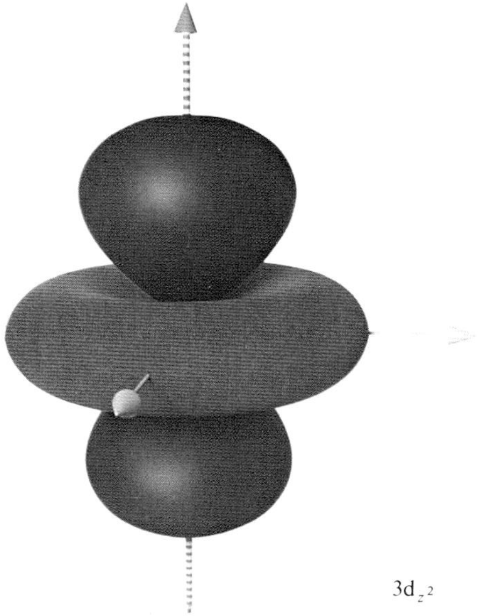 $3d_{z^2}$